earth

The poetry of earth is ceasing never.

—John Keats
"On the Grasshopper and Cricket"

Library of Congress Cataloging-in-Publication Data

Names: Carlowicz, Michael J., author. | Friedl, Lawrence, author. | Ward, Kevin A., 1968- author. | United States. National Aeronautics and Space Administration, issuing body.

Title: Earth / Michael Carlowicz, Lawrence Friedl, Kevin Ward.

Description: Washington, DC : National Aeronautics and Space Administration, Earth Science, NASA Headquarters, [2018]

Identifiers: LCCN 2018048628| ISBN 9781626830479 (hardback) | ISBN 1626830479 (hardback)

Subjects: LCSH: Earth (Planet)--Photographs from space. | Earth (Planet)--Remote-sensing images. | Earth (Planet)

Classification: LCC QB637 .C37 2018 | DDC 910.020222--dc23 | SUDOC NAS 1.2:EA 7/21

LC record available at https://lccn.loc.gov/2018048628

ISBN 978-1-62683-047-9

90000

9781626830479

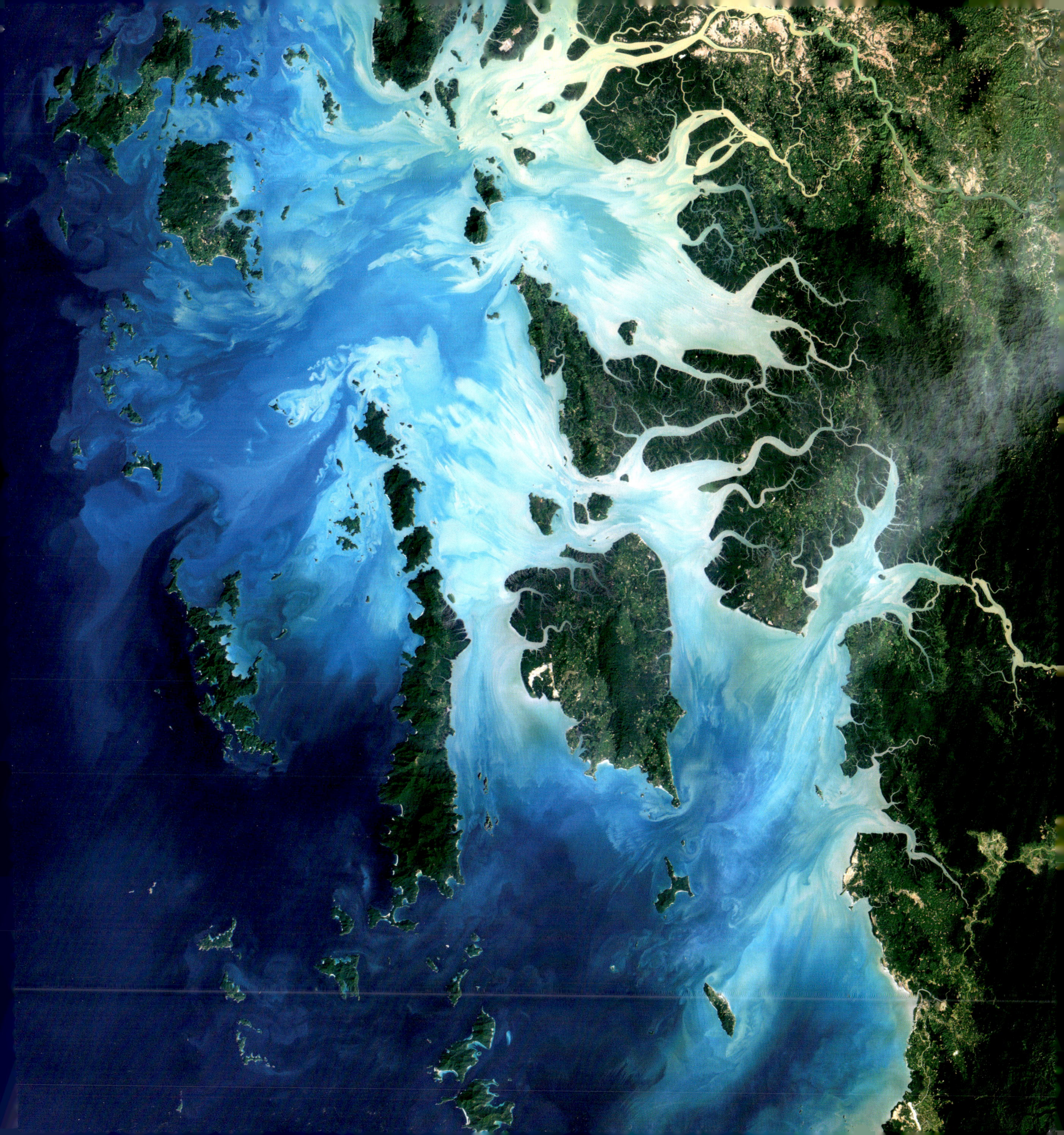

Contents

Foreword

Of all celestial bodies within reach or view, as far as we can see, out to the edge, the most wonderful and marvelous and mysterious is turning out to be our own planet Earth. There is nothing to match it anywhere, not yet anyway.

—Lewis Thomas

Sixty years ago, with the launch of Explorer 1, NASA made its first observations of Earth from space. Fifty years ago, astronauts left Earth orbit for the first time and looked back at our "blue marble." All of these years later, as we send spacecraft and point our telescopes past the outer edges of the solar system, as we study our planetary neighbors and our Sun in exquisite detail, there remains much to see and explore at home.

We are still just getting to know Earth through the tools of science. For centuries, painters, poets, philosophers, and photographers have sought to teach us something about our home through their art.

This book stands at an intersection of science and art. From its origins, NASA has studied our planet in novel ways, using ingenious tools to study physical processes at work—from beneath the crust to the edge of the atmosphere. We look at it in macrocosm and microcosm, from the flow of one mountain stream to the flow of jet streams. Most of all, we look at Earth as a system, examining the cycles and processes—the water cycle, the carbon cycle, ocean circulation, the movement of heat—that interact and influence each other in a complex, dynamic dance across seasons and decades.

We measure particles, gases, energy, and fluids moving in, on, and around Earth. And like artists, we study the light—how it bounces, reflects, refracts, and gets absorbed and changed. Understanding the light and the pictures it composes is no small feat, given the rivers of air and gas moving between our satellite eyes and the planet below.

For all of the dynamism and detail we can observe from orbit, sometimes it is worth stepping back and simply admiring Earth. It is a beautiful, awe-inspiring place, and it is the only world most of us will ever know.

NASA has a unique vantage point for observing the beauty and wonder of Earth and for making sense of it. Looking back from space, astronaut Edgar Mitchell once called Earth "a sparkling blue and white jewel," and it does dazzle the eye. The planet's palette of colors and textures and shapes—far more than just blues and whites—are spread across the pages of this book.

We chose these images because they inspire. They tell a story of a 4.5-billion-year-old planet where there is always something new to see. They tell a story of land, wind, water, ice, and air as they can only be viewed from above. They show us that no matter what the human mind can imagine, no matter what the artist can conceive, there are few things more fantastic and inspiring than the world as it already is. The truth of our planet is just as compelling as any fiction.

We hope you enjoy this satellite view of Earth.
It is your planet. It is NASA's mission.

Michael Carlowicz

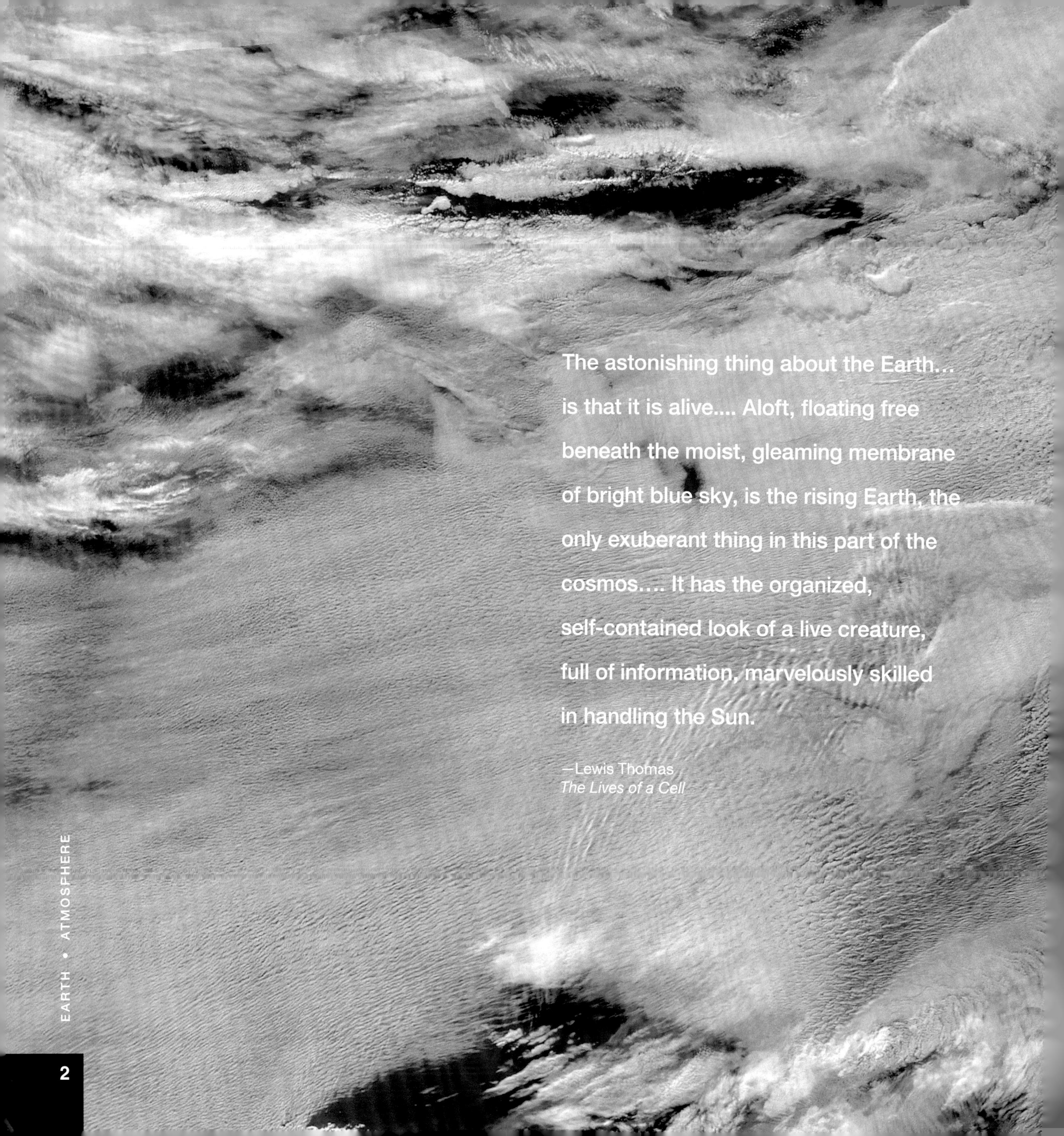

The astonishing thing about the Earth... is that it is alive.... Aloft, floating free beneath the moist, gleaming membrane of bright blue sky, is the rising Earth, the only exuberant thing in this part of the cosmos.... It has the organized, self-contained look of a live creature, full of information, marvelously skilled in handling the Sun.

—Lewis Thomas
The Lives of a Cell

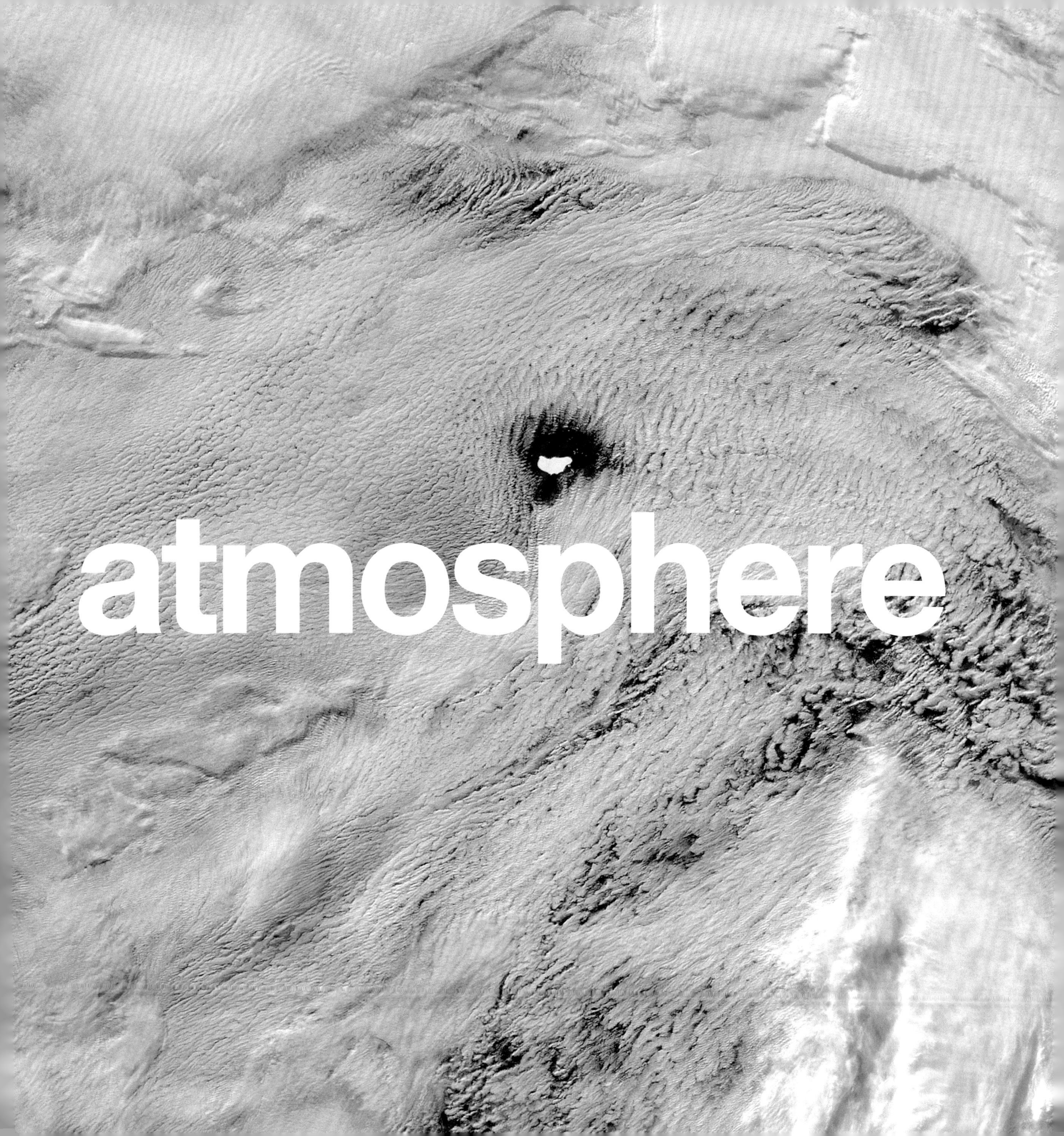
atmosphere

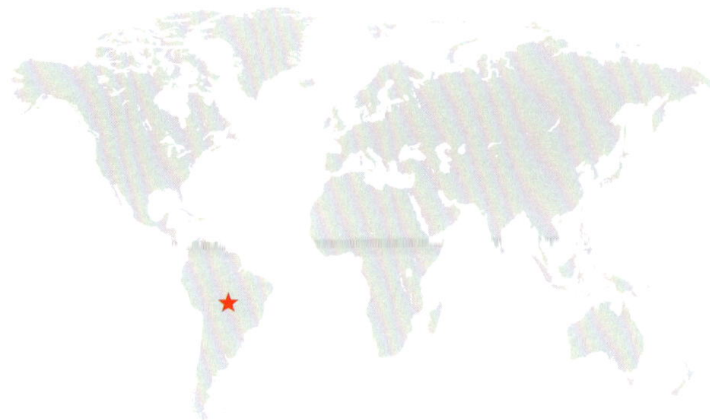

Curving Cloud Streets
Brazil and Bolivia

To the human eye, the wind is invisible. It can only be visualized by proxy, by its expressions in other phenomena like blowing leaves, airborne dust, white-capped waters—or the patterns of clouds.

Acquired in June 2014 by the Aqua satellite, this image shows a broad swath of the Amazon rainforest in Brazil and Bolivia as it appeared in the early afternoon. As sunlight warms the forest in the morning, water vapor rises on columns of heated air. When that humid air runs into a cooler, more stable air mass above, it condenses into fluffy cumulus clouds.

Cumulus cloud streets often trace the direction, and sometimes the intensity, of winds—lining up parallel to the direction of the wind. Usually this means a straight line, but clouds can also line up along the concentric, curved lines of high-pressure weather systems, as they did here.

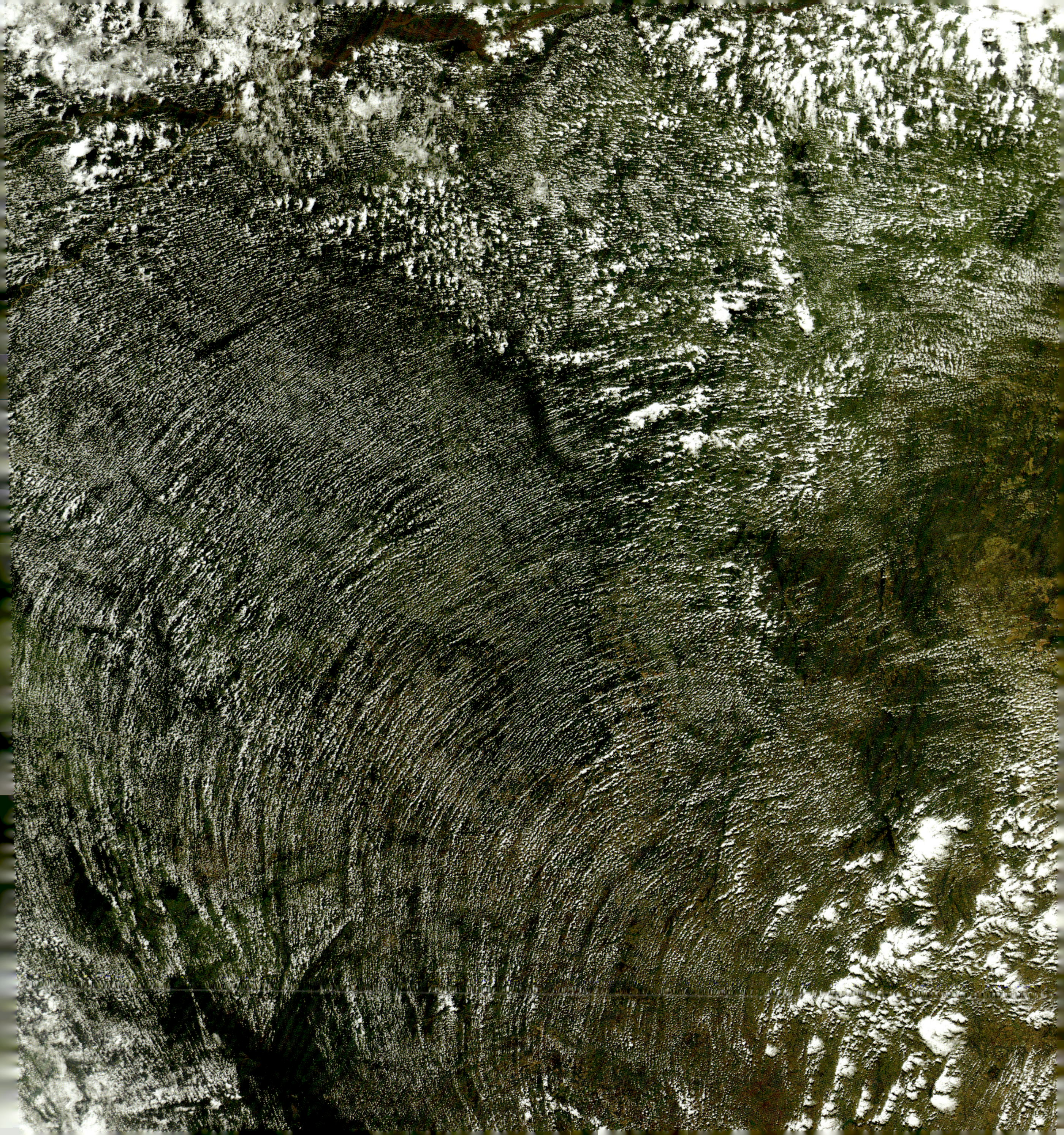

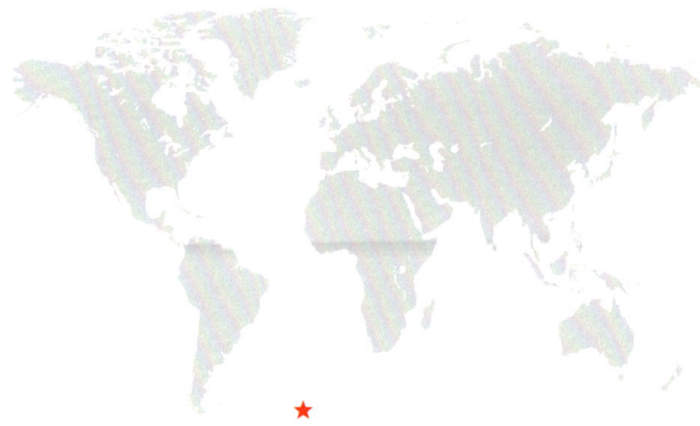

A Trio of Plumes
South Atlantic Ocean

The uninhabited South Sandwich Islands include several active stratovolcanoes. Due to their remote location, these volcanoes are some of the least studied in the world, though satellites often catch them erupting.

The combination of clouds and ice at these latitudes can make it difficult to see plumes of volcanic ash in natural-color imagery. But using portions of the electromagnetic spectrum that are typically invisible to the naked eye (such as infrared) enables satellites to distinguish ice from ash and clouds. The Aqua satellite captured this false-color image in September 2016. Note the three bright white plumes running down the middle third of the page; they are warmer and brighter in infrared than the cooler ice clouds (teal) around them.

Researchers have learned that even small eruptions like this can affect cloud cover and weather. The tiny solid and liquid particles in the plume (aerosols) act as seeds for the formation of cloud droplets.

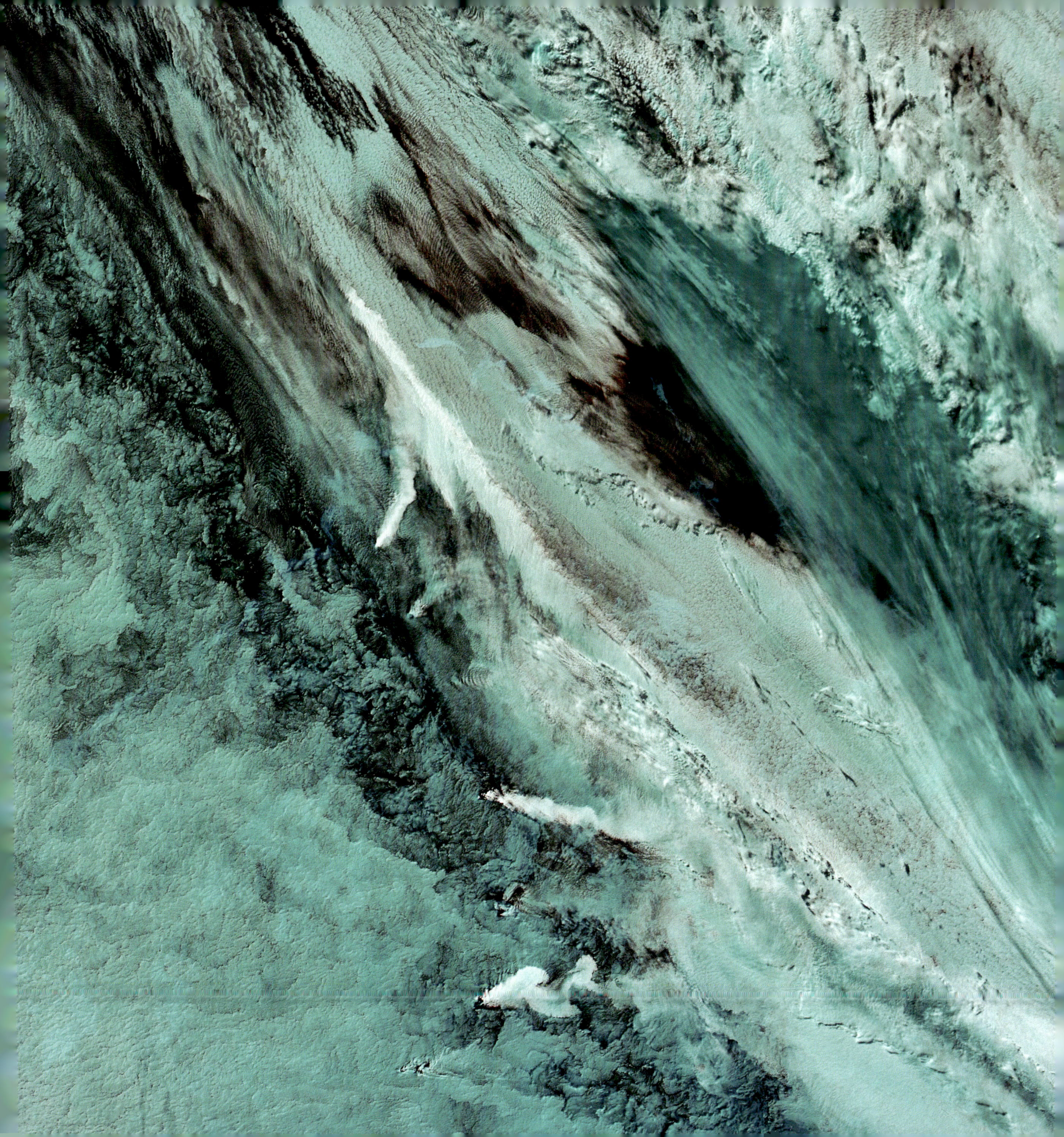

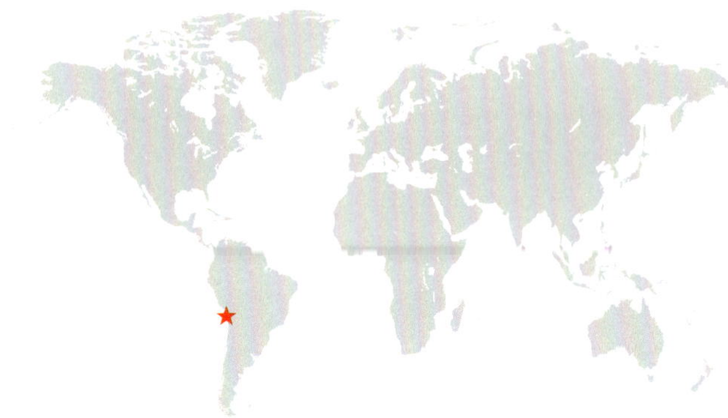

Filling the Valleys
Peru

The valleys along Peru's southern coast are among the deepest on Earth. They are also frequently filled with clouds. In July 2015, Landsat 8 captured this view of the cloud-filled canyons through which the Yauca and Acarí rivers empty into the Pacific Ocean.

You can't see it, but the Pacific lies below the clouds on the lower left. The clouds are marine stratocumulus—a type of low-level cloud so close to the surface that it is essentially fog. Such clouds are a persistent feature off the coast of Peru and Chile, developing most often during the winter and early spring. On some occasions, prevailing winds can push the clouds inland. Because the marine clouds are low, they are easily blocked by coastal mountains and hills, such as the Andes. But in areas where low valleys open to the ocean, the clouds move inland.

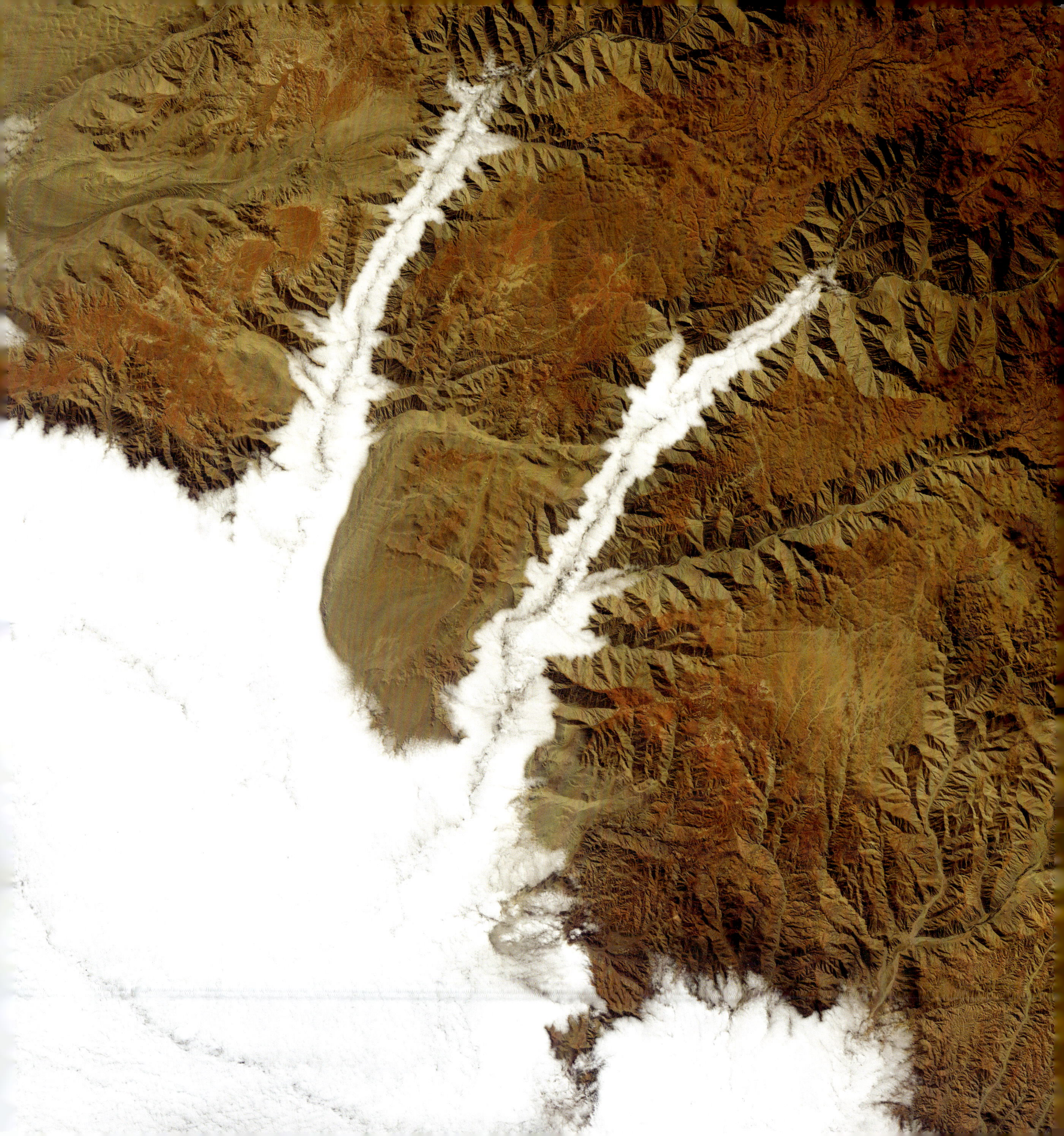

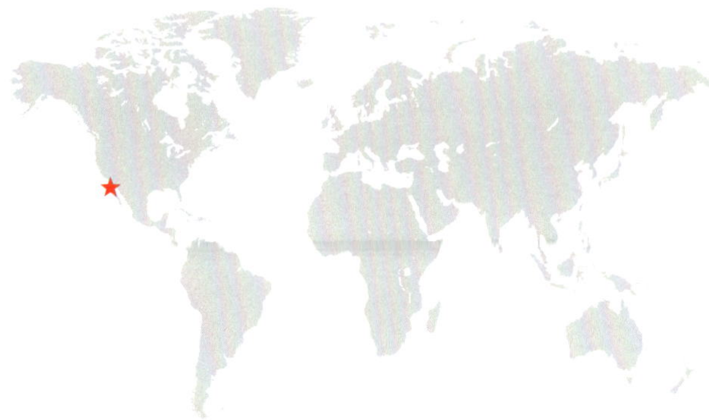

A Glorious View
Pacific Ocean

A layer of stratocumulus clouds over the Pacific Ocean serves as the backdrop for this rainbow-like phenomenon known as a glory. Glories form when water droplets within clouds scatter sunlight back toward a source of illumination (in this case, the Sun).

Although glories may look similar to rainbows, the way light is scattered to produce them is different. Rainbows are formed by refraction and reflection; glories are formed by backward diffraction. From the ground or from an airplane, glories appear as circular rings of color. In this image, however, the glory is stretched vertically because of how the imager scans the surface in swaths.

Note, too, the swirling von Kármán vortices visible to the right of the glory. The alternating rows of vortices form as air masses run into an obstacle—the island of Guadalupe—and form a wake behind it.

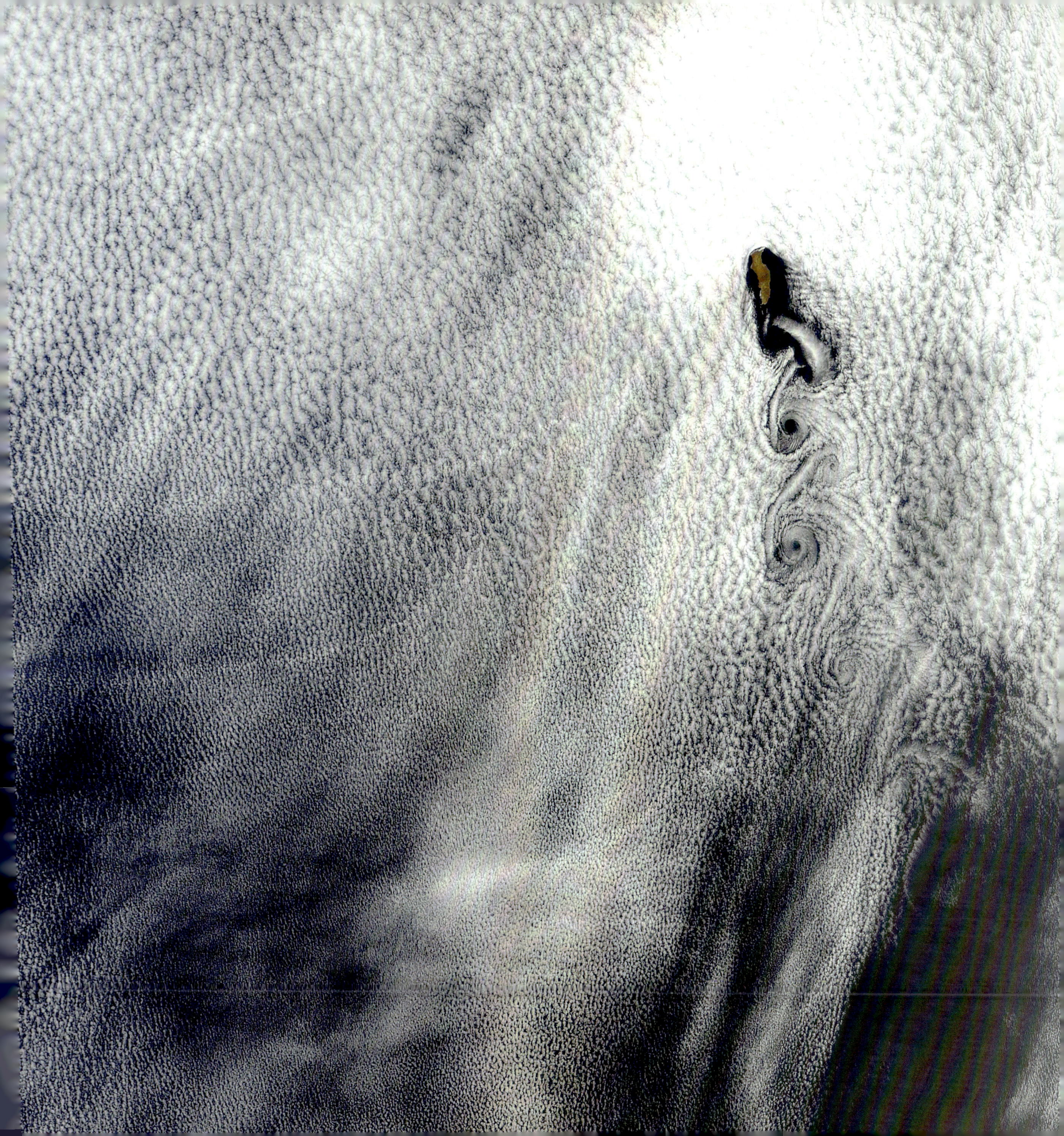

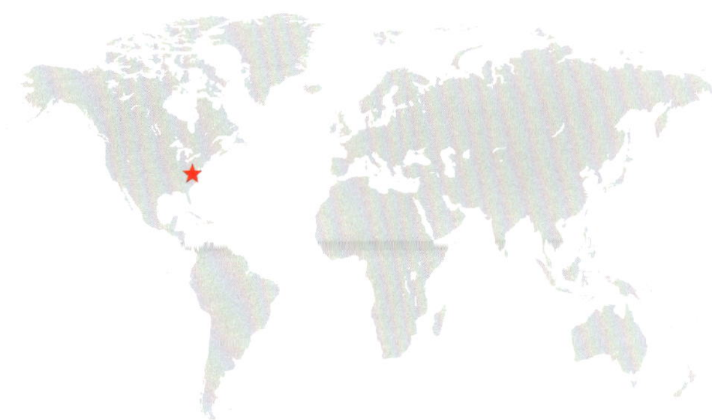

Punching Holes in the Sky
United States

In December 2009, the Landsat 5 satellite observed this extraordinary example of "hole-punch clouds" over West Virginia. This strange phenomenon results from a combination of cold temperatures, air traffic, and atmospheric instability.

If you were to look from below, it would appear as if part of the cloud were falling out of the sky. As it turns out, that is actually what is happening. The clouds are initially composed of liquid drops at a super-cooled temperature below 0° Celsius. As an airplane passes through a cloud, particles in the exhaust can create a disturbance that triggers freezing. Ice particles then quickly grow at the expense of water droplets. Eventually, the ice crystals in these patches of clouds grow large enough that they literally fall out of the sky, earning hole-punch clouds their alternate name: "fallstreak holes."

In this false-color image, pink and faint blue areas are typical water-rich clouds, and bright cyan areas are ice clouds. The hole-punch areas appear dark around the ice clouds.

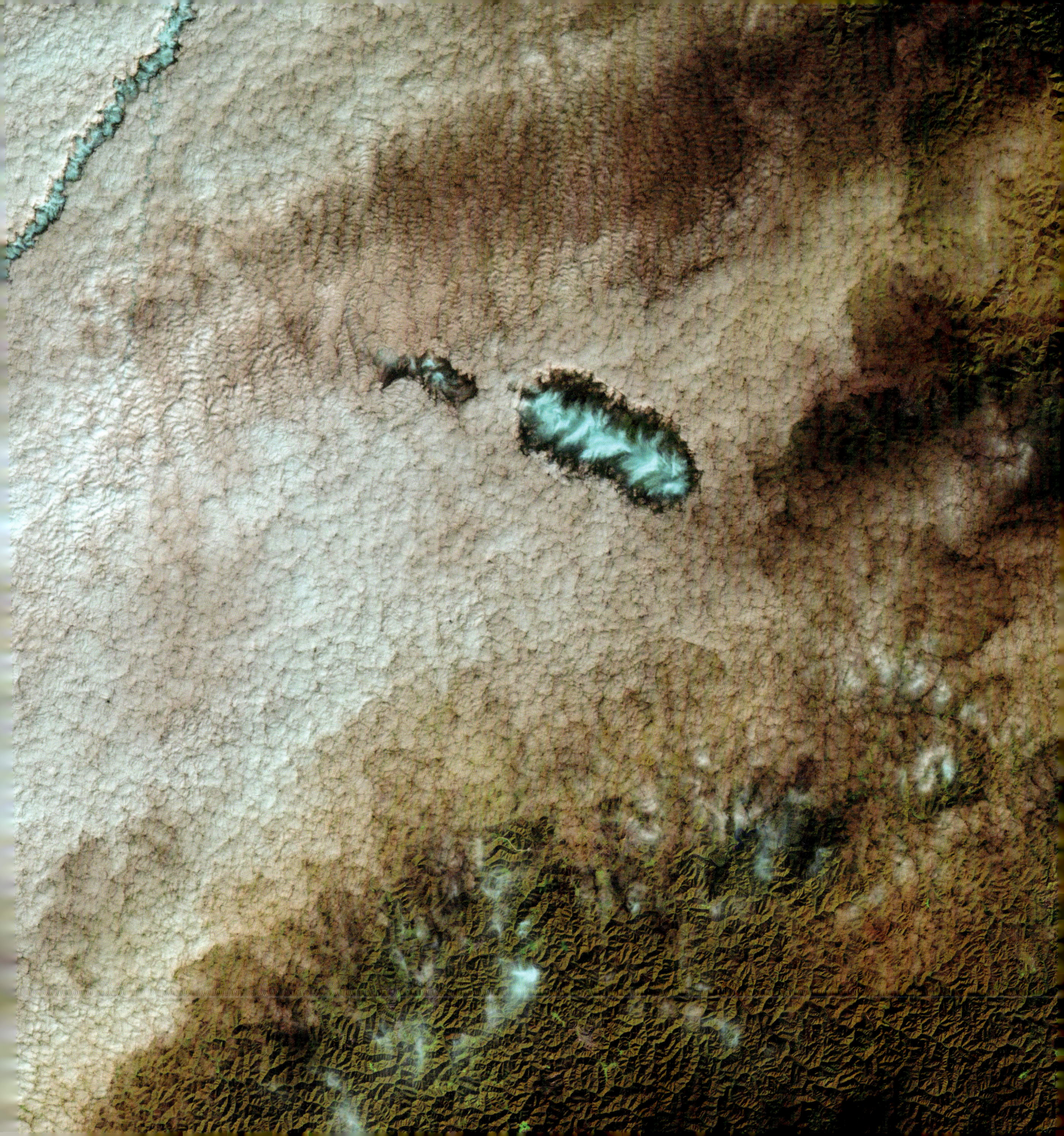

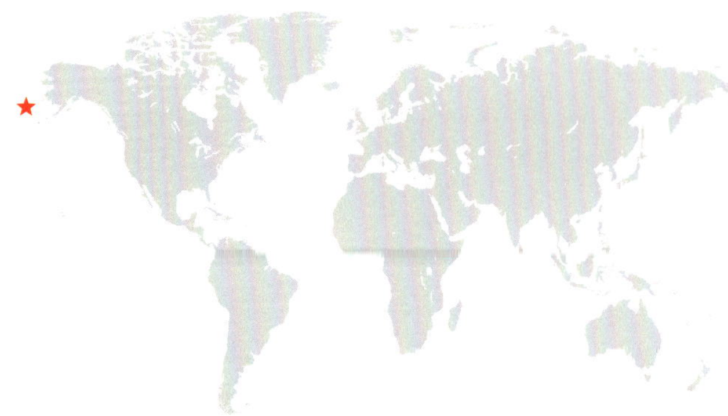

Bering Streets
Arctic Ocean

Winds from the northeast pushed sea ice southward and formed cloud streets—parallel rows of clouds—over the Bering Strait in January 2010. The easternmost reaches of Russia, blanketed in snow and ice, appear in the upper left. To the east, sea ice spans the Bering Strait. Along the southern edge of the ice, wavy tendrils of newly formed, thin sea ice predominate.

The cloud streets run in the direction of the northerly wind that helps form them. When wind blows out from a cold surface like sea ice over the warmer, moister air near the open ocean, cylinders of spinning air may develop. Clouds form along the upward cycle in the cylinders, where air is rising, and skies remain clear along the downward cycle, where air is falling. The cloud streets run toward the southwest in this image from the Terra satellite.

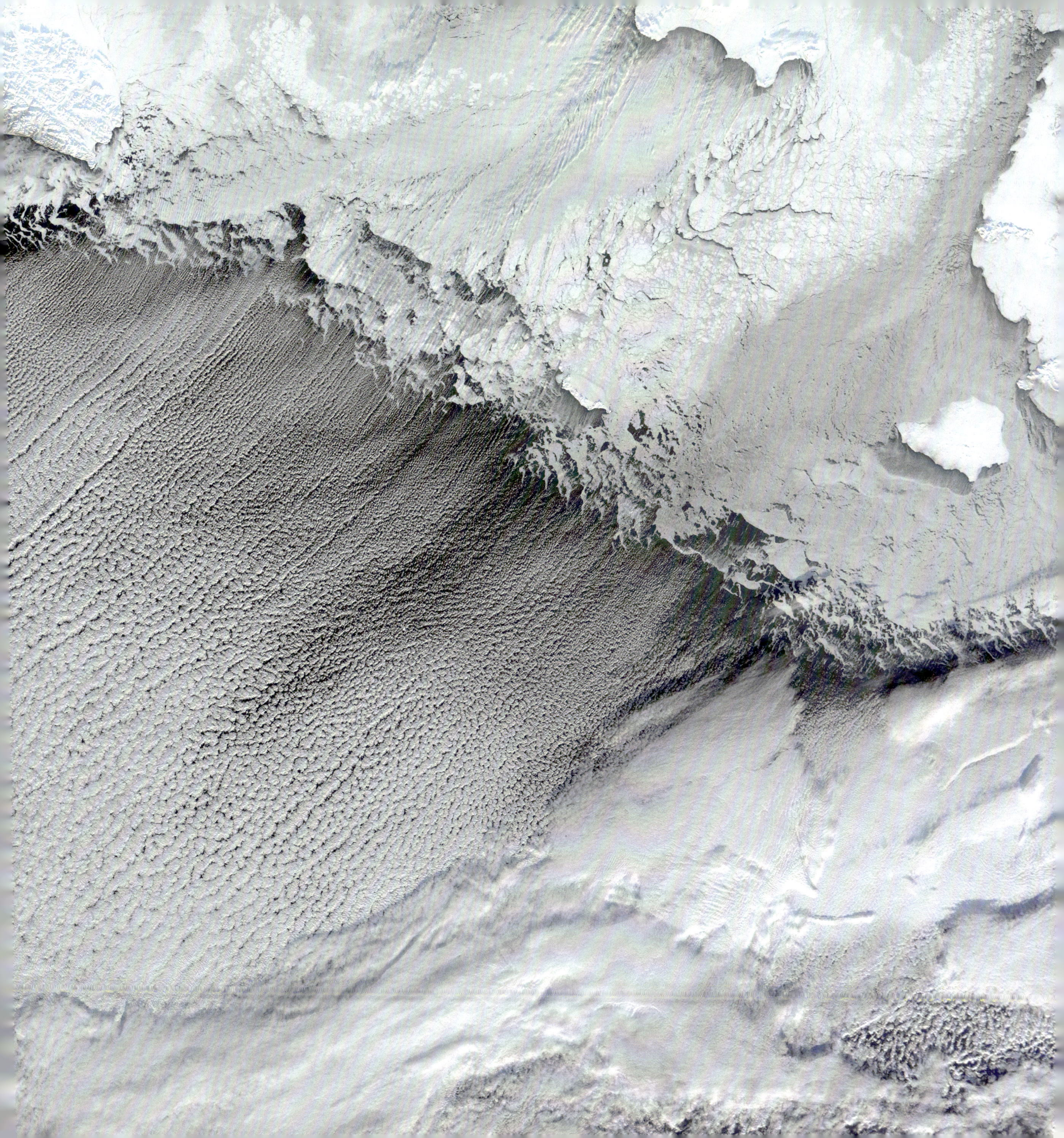

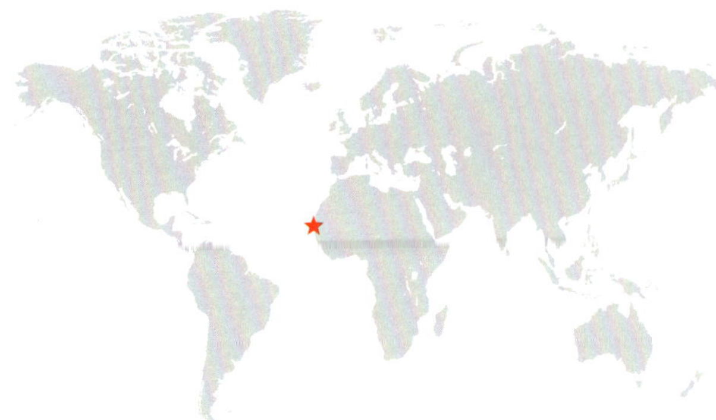

Riding the Waves
Mauritania

You cannot see it directly, but air masses from Africa and the Atlantic Ocean are colliding in this Landsat 8 image from August 2016. The collision off the coast of Mauritania produces a wave structure in the atmosphere.

Called an undular bore or solitary wave, this cloud formation was created by the interaction between cool, dry air coming off the continent and running into warm, moist air over the ocean. The winds blowing out from the land push a wave of air ahead like a bow wave moving ahead of a boat.

Parts of these waves are favorable for cloud formation, while other parts are not. The dust blowing out from Africa appears to be riding these waves. Dust has been known to affect cloud growth, but it probably has little to do with the cloud pattern observed here.

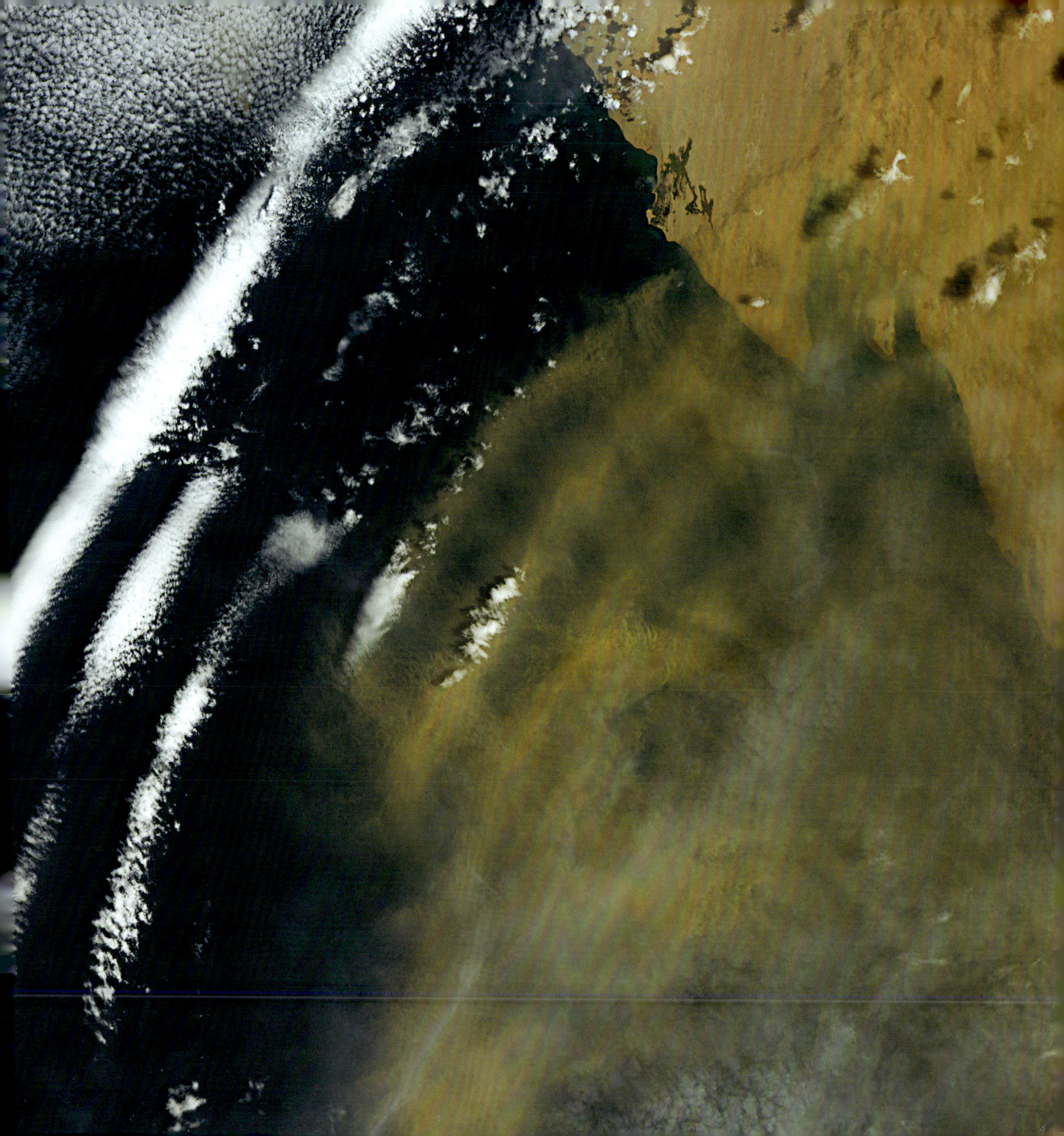

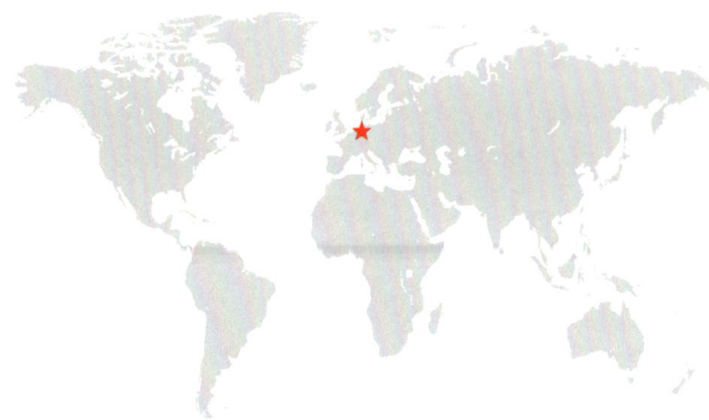

Cloud Shadow
Germany

In November 2012, the Earth Observing-1 satellite acquired this image of a layer of clouds casting a distinctive shadow on another, lower cloud layer. The upper deck was more than 1 kilometer (0.6 miles) above the ground. Both layers were composed of stratus clouds, a low-lying type that tends to be uniform and flat. When the satellite's Advanced Land Imager acquired the image, the clouds were over northeastern Germany near Harz National Park.

While local meteorological conditions affect cloud height, the latitude at which clouds form is also important. Almost all clouds exist in the lowest level of the atmosphere, the troposphere. However, the depth of the troposphere varies by latitude—thinner near the poles than the Equator, so clouds can occur at higher levels in the tropics than they do at high- and mid-latitudes.

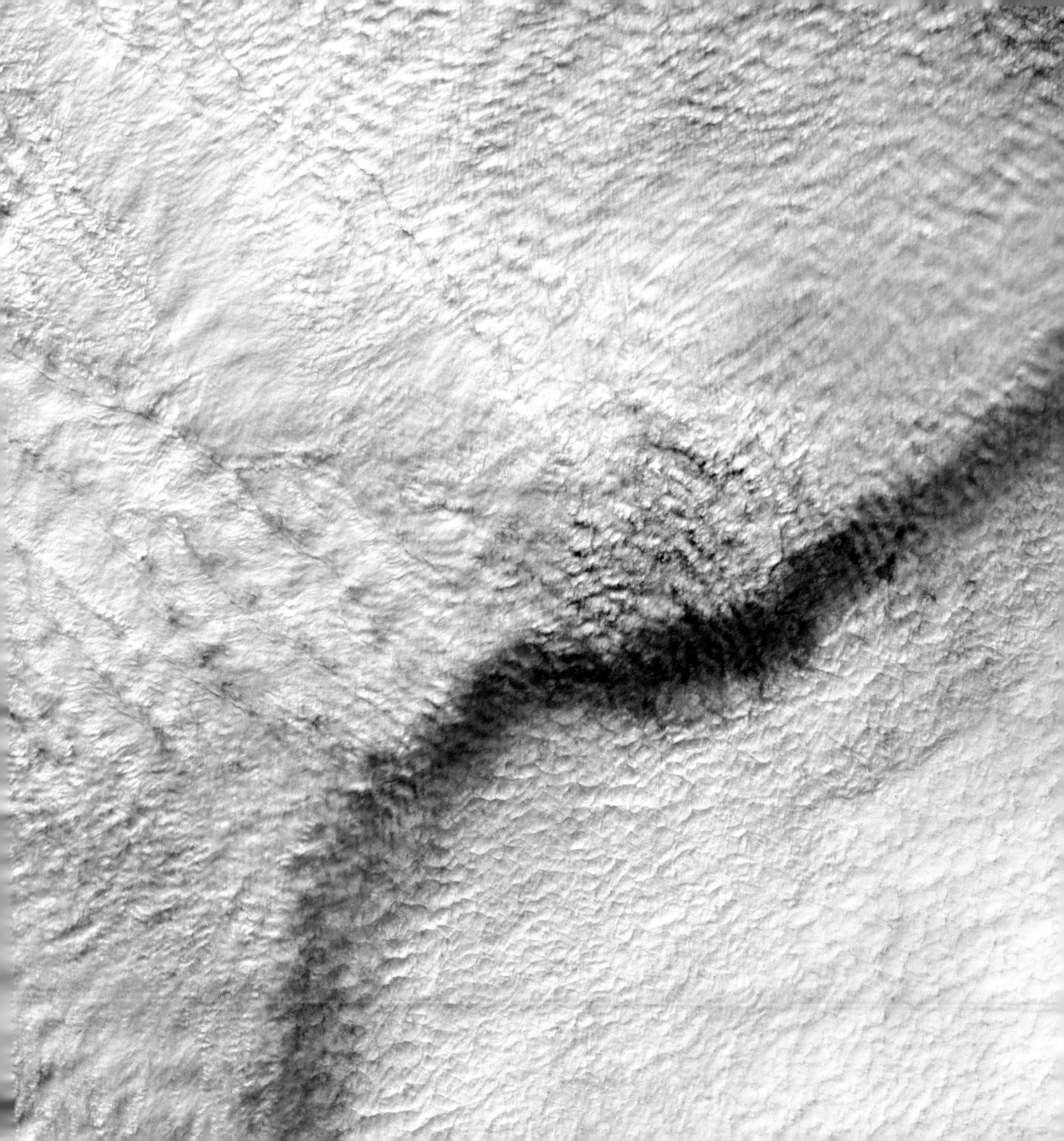

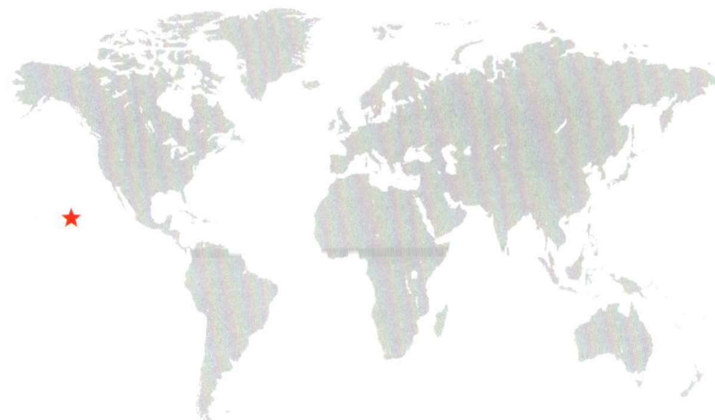

Double Trouble
Pacific Ocean

The island of Hawaii rarely takes a direct hit from a hurricane. In August 2016, two Pacific storms almost changed that.

The Suomi NPP satellite observed Hurricanes Madeline and Lester stirring up the central Pacific Ocean as category 3 and 4 storms while moving northwest toward the Hawaiian Islands. The tight, deep eye of category 3 Hurricane Madeline (right) appears almost three-dimensional even in a two-dimensional satellite view. Lester (next page) showed off an equally menacing eye. The bright streaks across the ocean surface are areas of sunglint, where sunlight reflected directly back at the image.

Ultimately, the storms blew just south and north of the islands without making landfall. In fact, no hurricane has made landfall on the Big Island since recordkeeping started in 1949. Only 15 hurricanes have passed within 200 nautical miles of the island in that time.

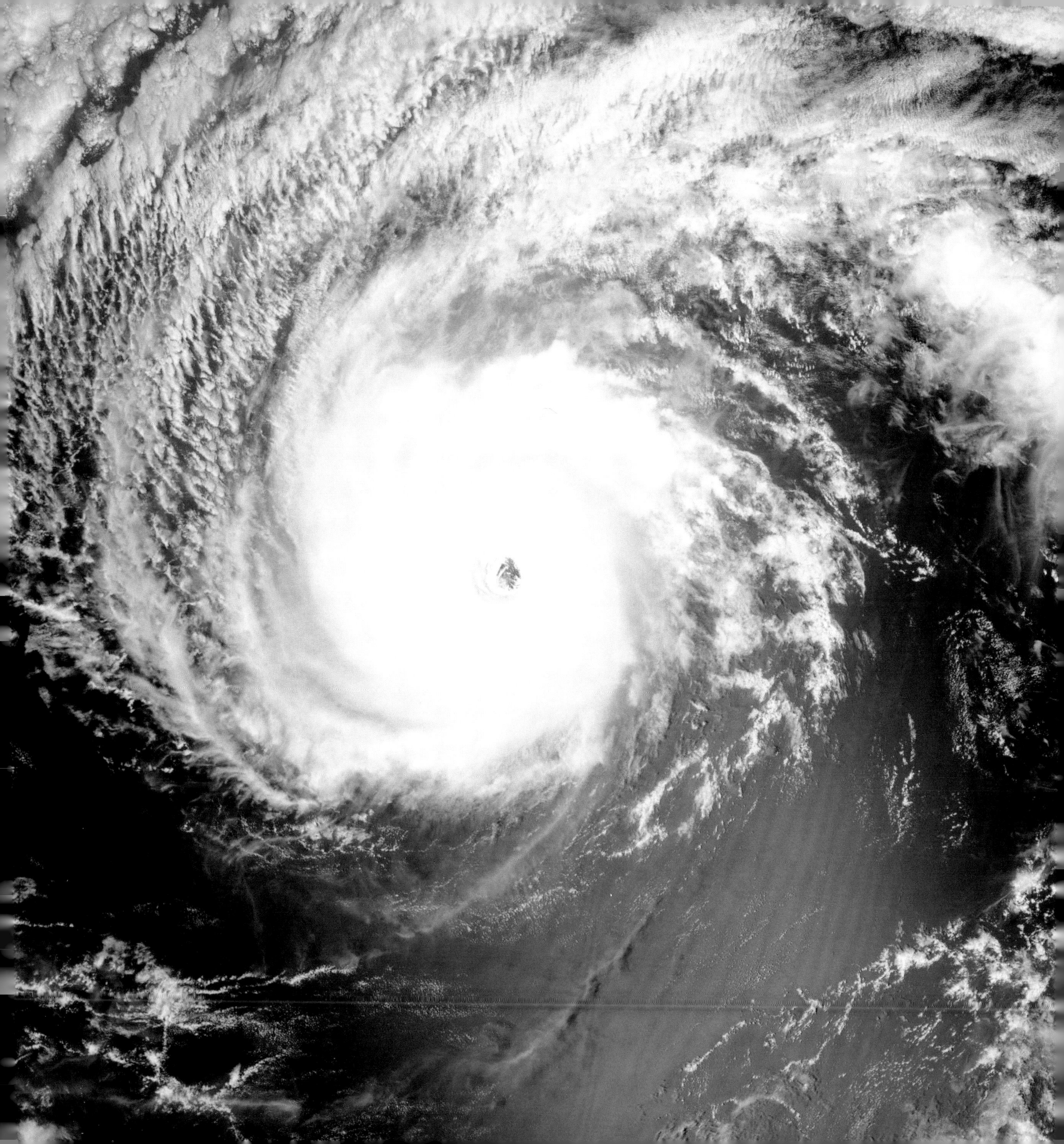

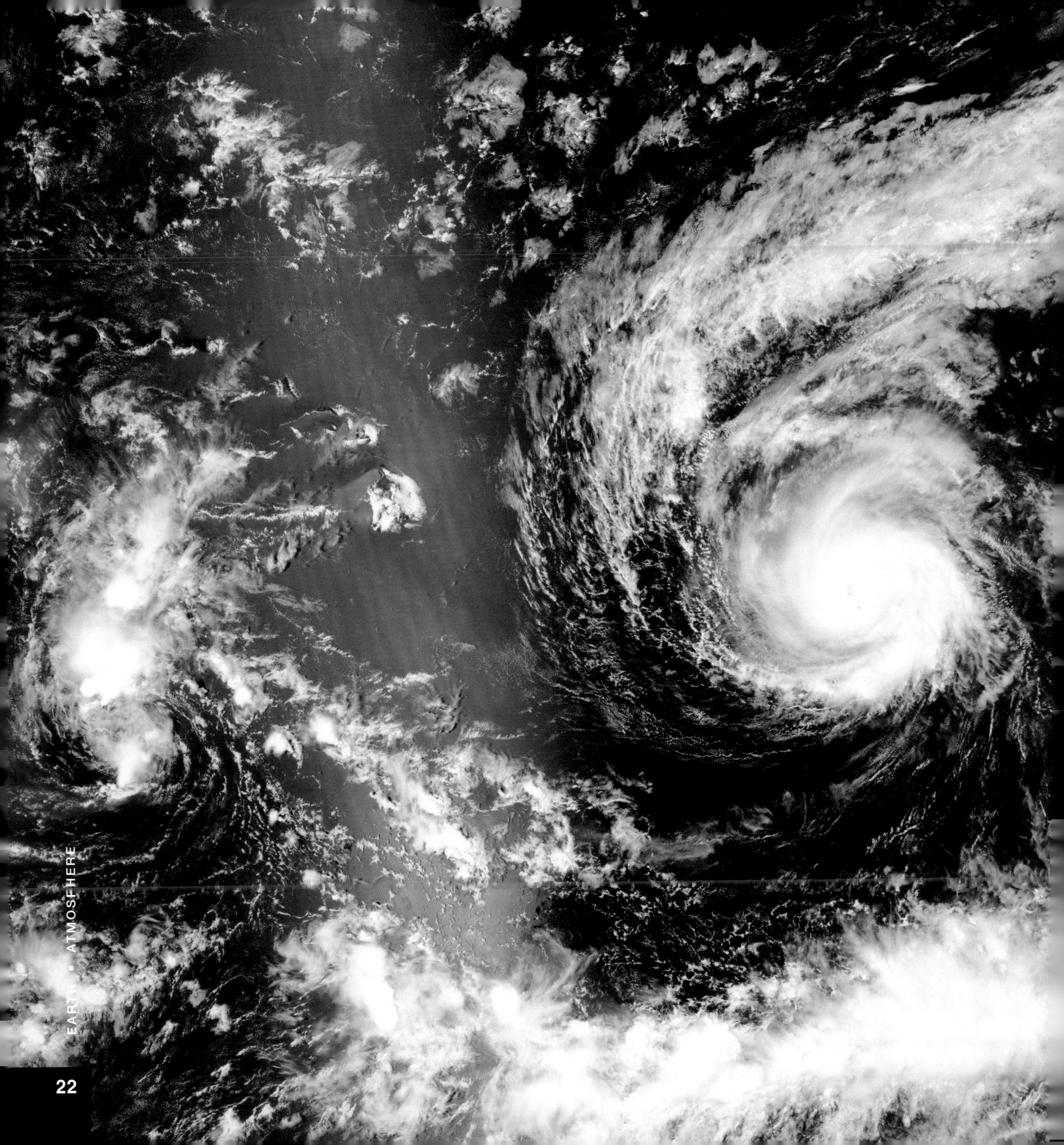

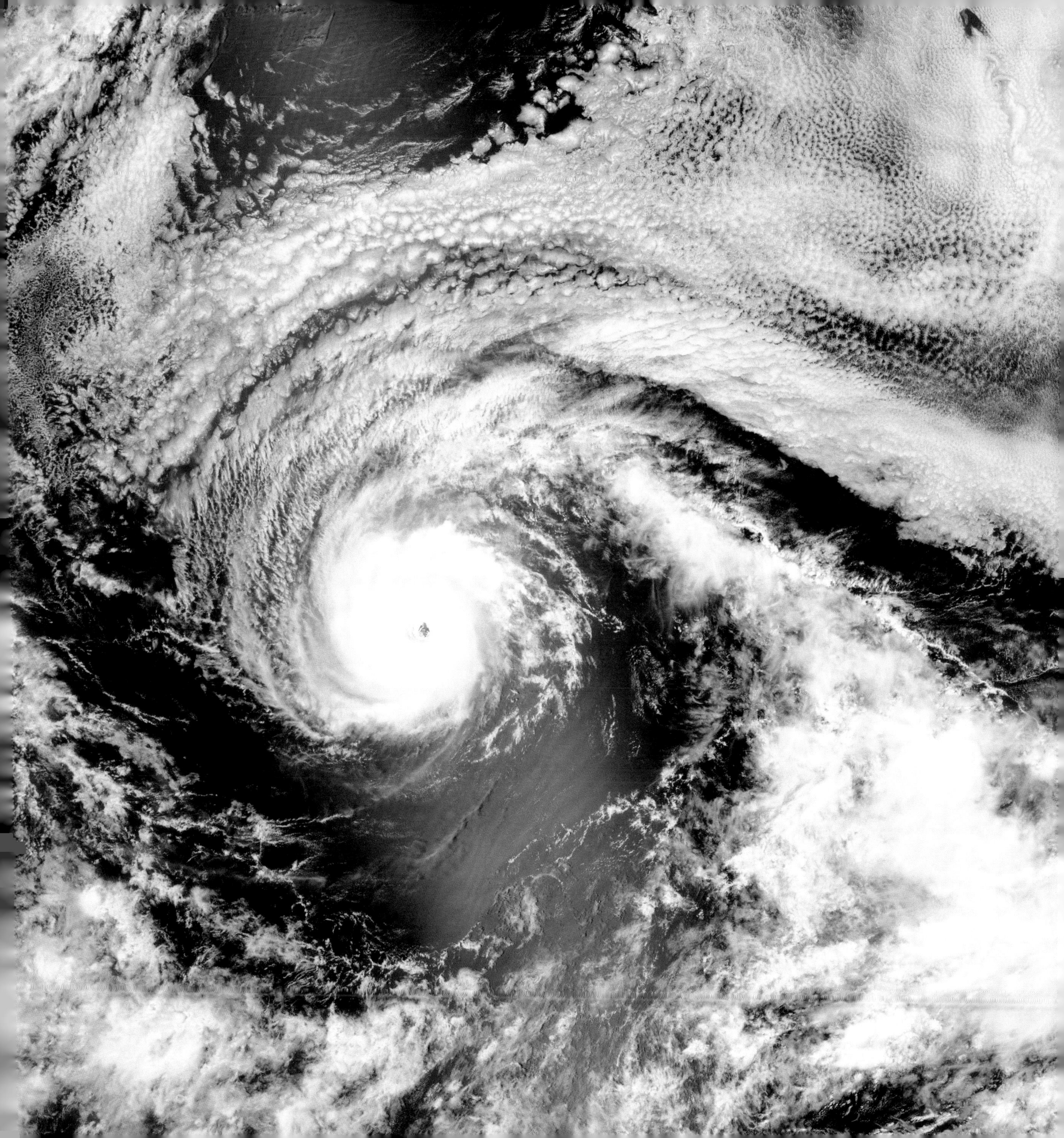

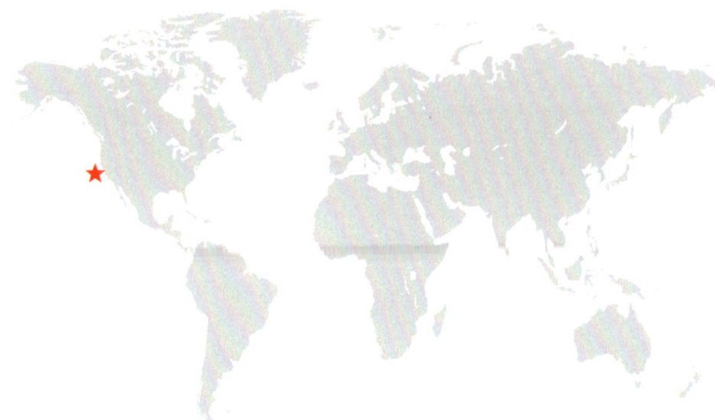

Making Tracks
Pacific Ocean

Ships steaming across the Pacific Ocean left this cluster of bright cloud trails lingering in the atmosphere in February 2012. The narrow clouds, known as ship tracks, form when water vapor condenses around tiny particles of pollution from ship exhaust. The crisscrossing clouds off the coast of California stretched for many hundreds of kilometers from end to end. The narrow ends of the clouds are youngest, while the broader, wavier ends are older.

Some of the pollution particles generated by ships (especially sulfates) are soluble in water and can serve as the seeds around which cloud droplets form. Clouds infused with ship exhaust have more and smaller droplets than unpolluted clouds. As a result, light hitting the ship tracks scatters in many directions, often making them appear brighter than other types of marine clouds, which are usually seeded by larger, naturally occurring particles like sea salt.

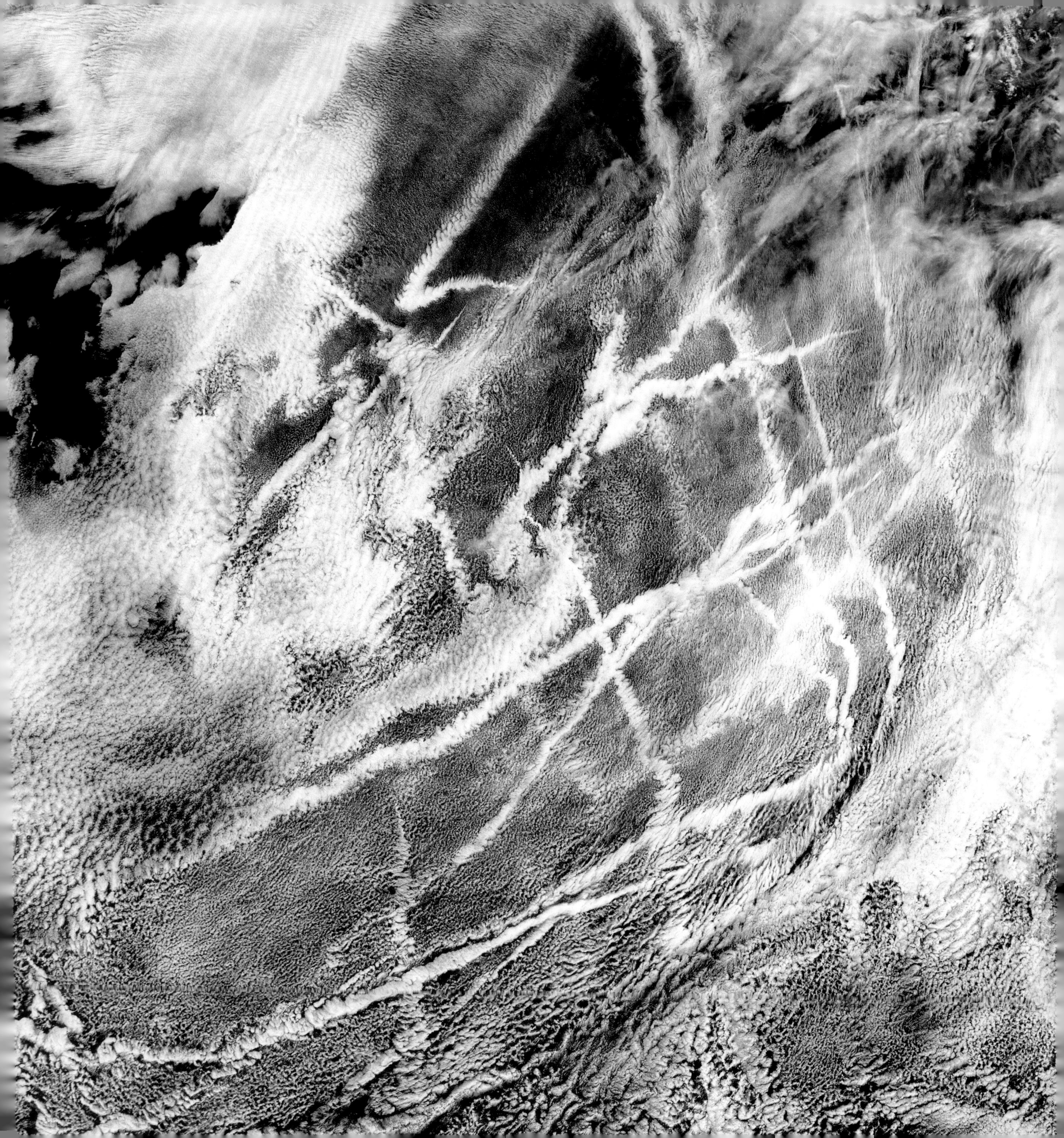

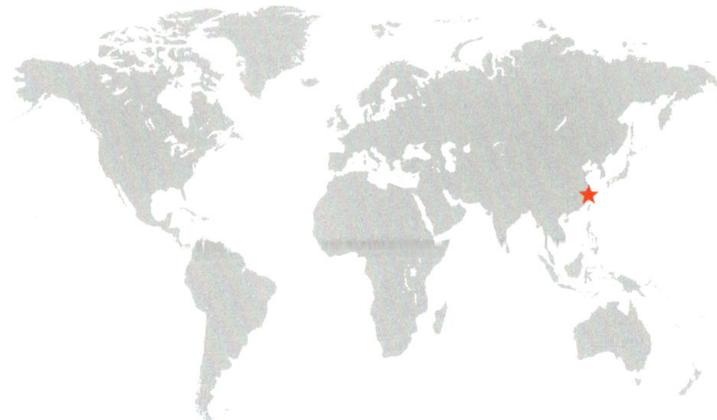

Tracing the Coast
China

The ocean does not heat up as much throughout a day as landmasses do. For this reason, cool, moist marine air commonly gives rise to dense clouds over the ocean.

In this example from the coast of China, an onshore wind carries the clouds from the ocean toward the land. But the land is warmer, drier, and unfavorable for cloud growth. As a result, the marine clouds that move onshore tend to evaporate, leaving a cloud layer that traces the coastline.

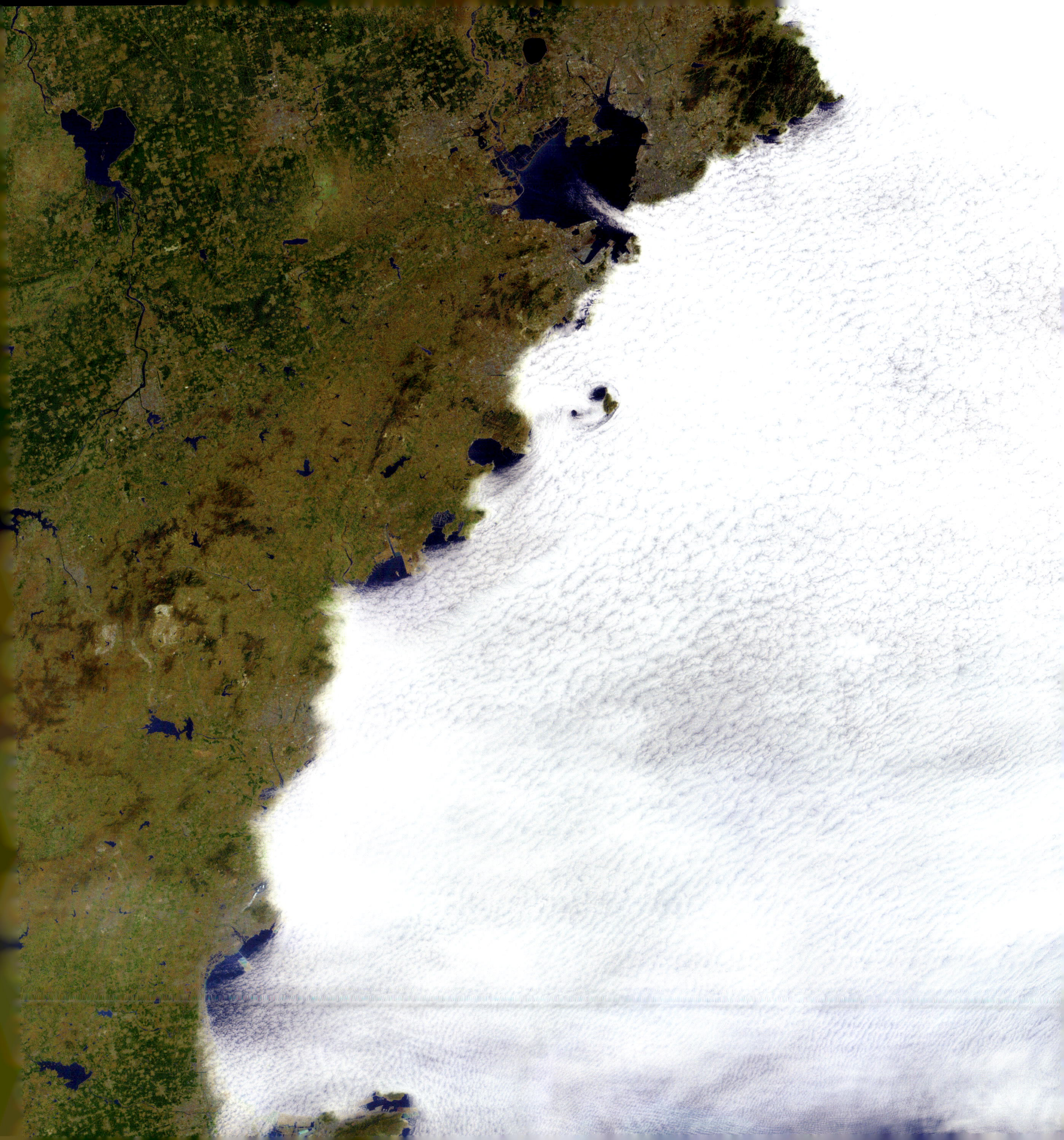

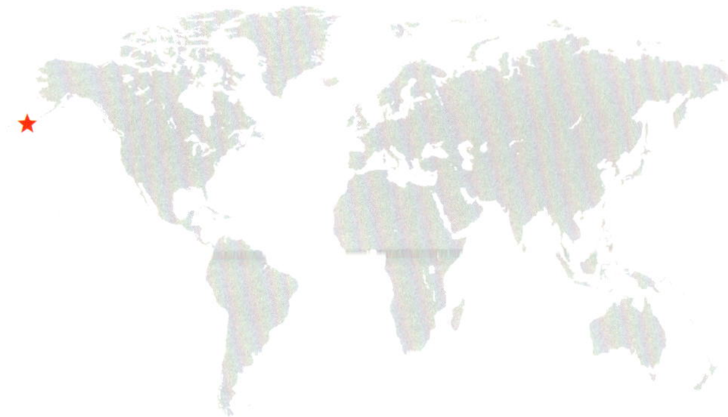

Four Mountains Stand Out
Pacific Ocean

They are called the Islands of the Four Mountains. Part of the Aleutian Island chain, these peaks are actually the upper slopes of volcanoes rising from the seafloor: Carlisle, Cleveland, Herbert, and Tana. Standing in one of the most remote reaches of the world, these volcanoes have scarcely been studied. Satellite sensing makes that easier, as this Landsat 8 image from June 2013 shows.

Herbert Island (right) is dominated by a symmetrical stratovolcano that stands in its center. The remote island has scarcely been studied, and there are no records of eruptions there. The 2-kilometer-wide summit caldera include a lake of meltwater, remnants of the snow that covers the peak for most of the year. The straight-down (nadir) satellite view can make it difficult to determine which part of the landscape stands taller than the other, a phenomenon known as relief inversion.

On the next page, you can see how a layer of low clouds and fog obscures the lower elevations of the islands and the sea surface. But these clouds also hint at the complicated airflow patterns around and through the islands.

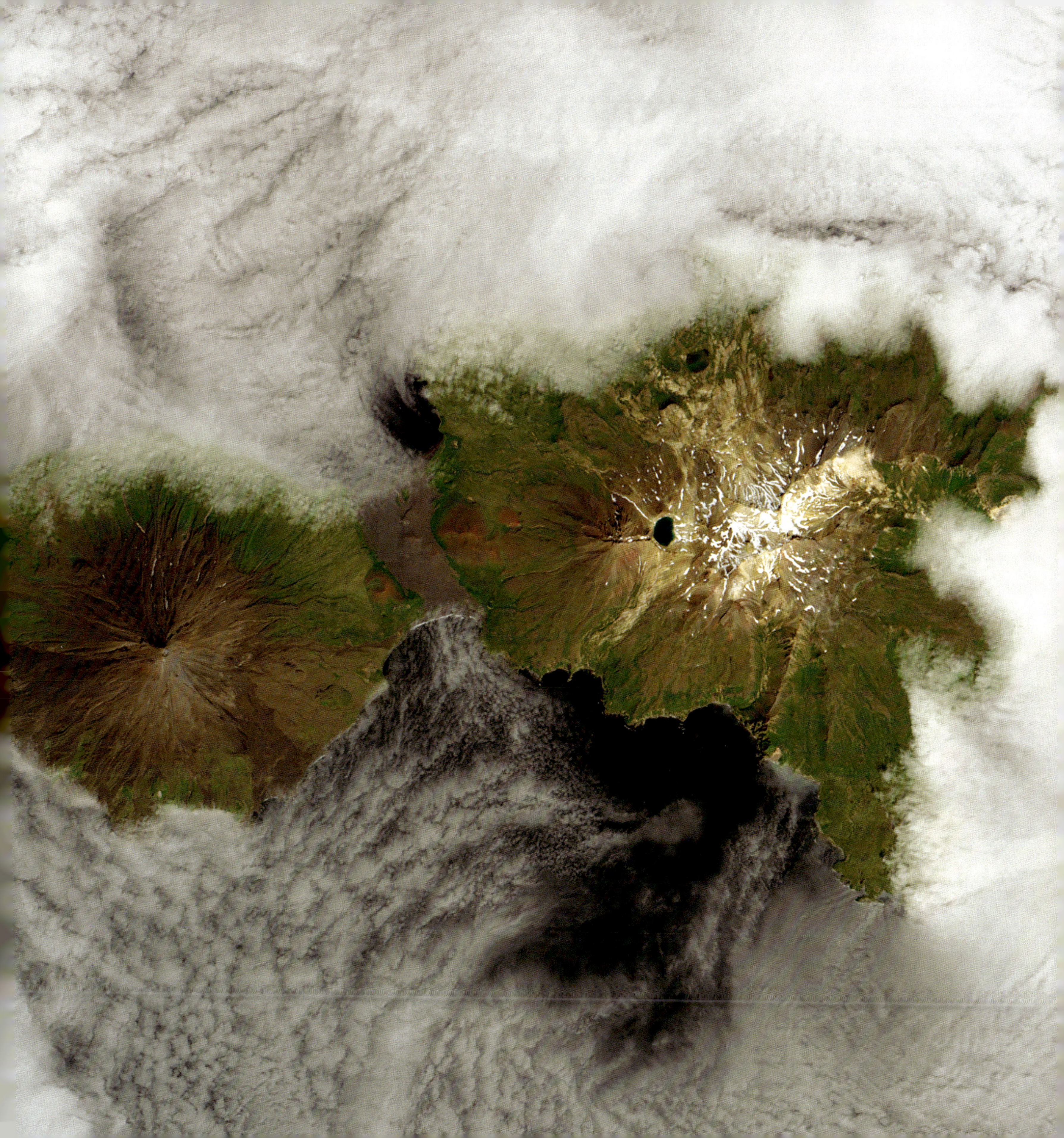

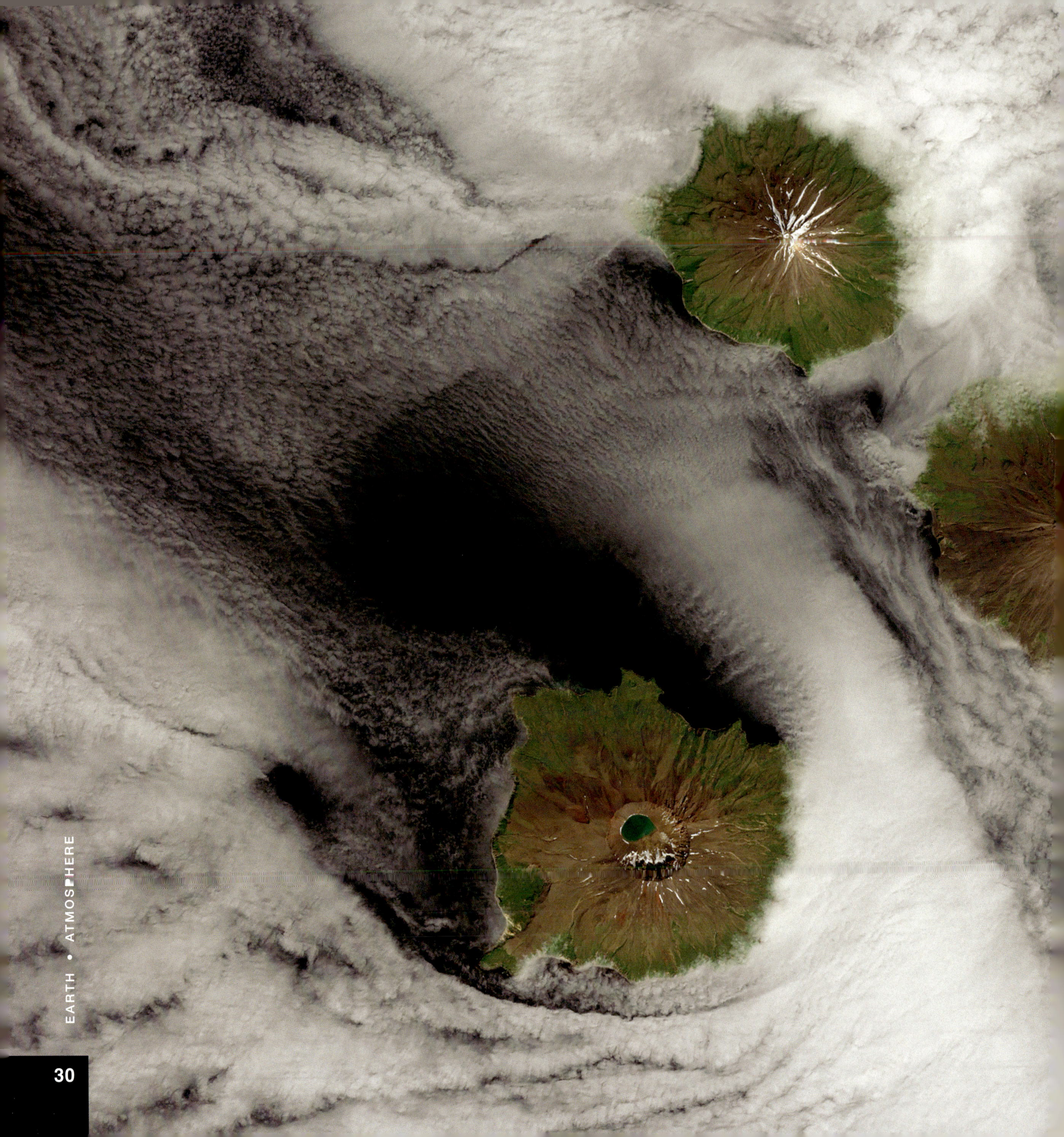

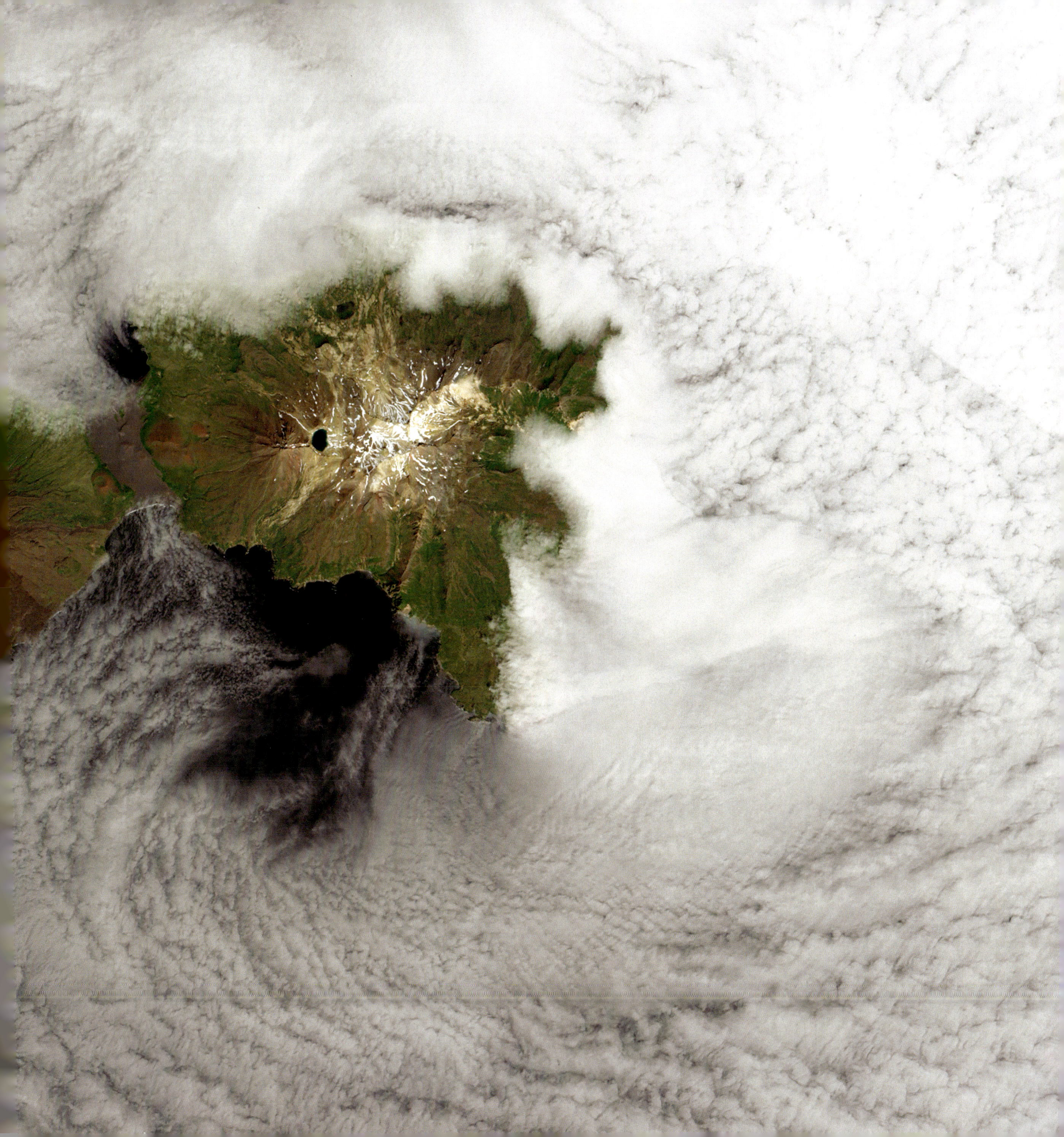

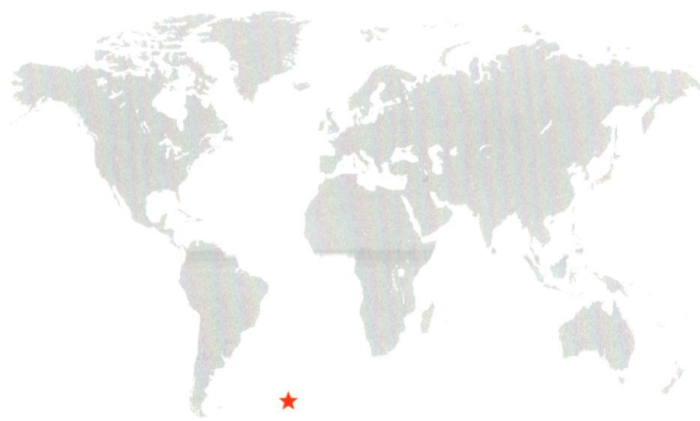

Framing an Iceberg South Atlantic Ocean

In June 2016, the Suomi NPP satellite captured this image of various cloud formations in the South Atlantic Ocean. Note how low stratus clouds framed a hole over iceberg A-56 as it drifted across the sea.

The exact reason for the hole in the clouds is somewhat of a mystery. It could have formed by chance, although imagery from the days before and after this date suggest something else was at work. It could be that the relatively unobstructed path of the clouds over the ocean surface was interrupted by thermal instability created by the iceberg. In other words, if an obstacle is big enough, it can divert the low-level atmospheric flow of air around it, a phenomenon often caused by islands.

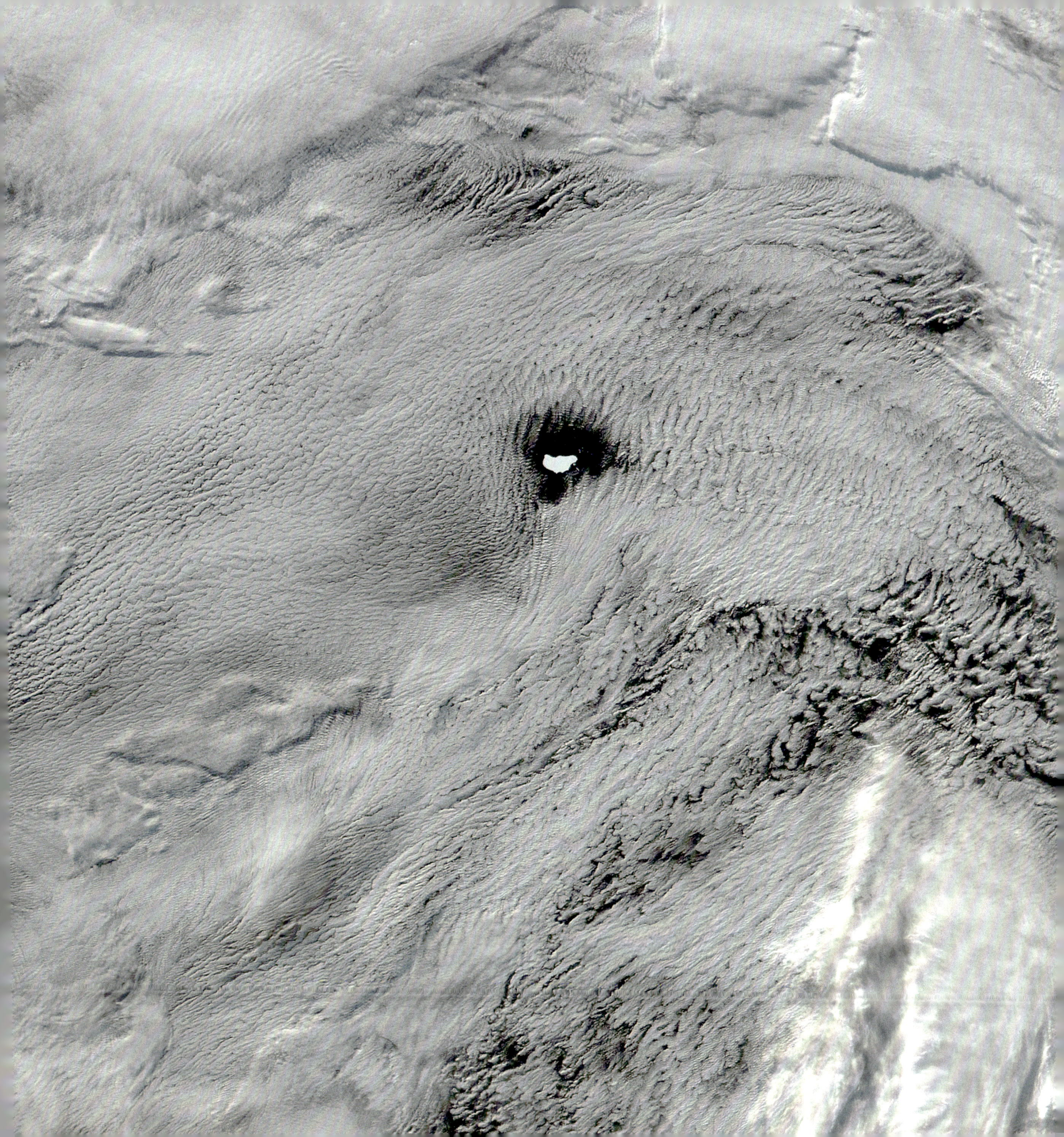

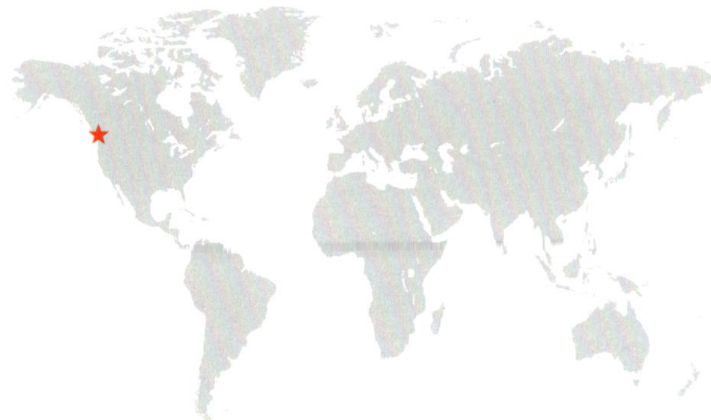

Valley Fog
Canada

Fog is essentially a cloud lying on the ground. Like all clouds, it forms when the air reaches its dew point—the temperature at which an air mass is cool enough for its water vapor to condense into liquid droplets.

This false-color image shows valley fog, which is common in the Pacific Northwest of North America. On clear winter nights, the ground and overlying air cool off rapidly, especially at high elevations. Cold air is denser than warm air, and it sinks down into the valleys. The moist air in the valleys gets chilled to its dew point, and fog forms. If undisturbed by winds, such fog may persist for days. The Terra satellite captured this image of foggy valleys northeast of Vancouver in February 2010.

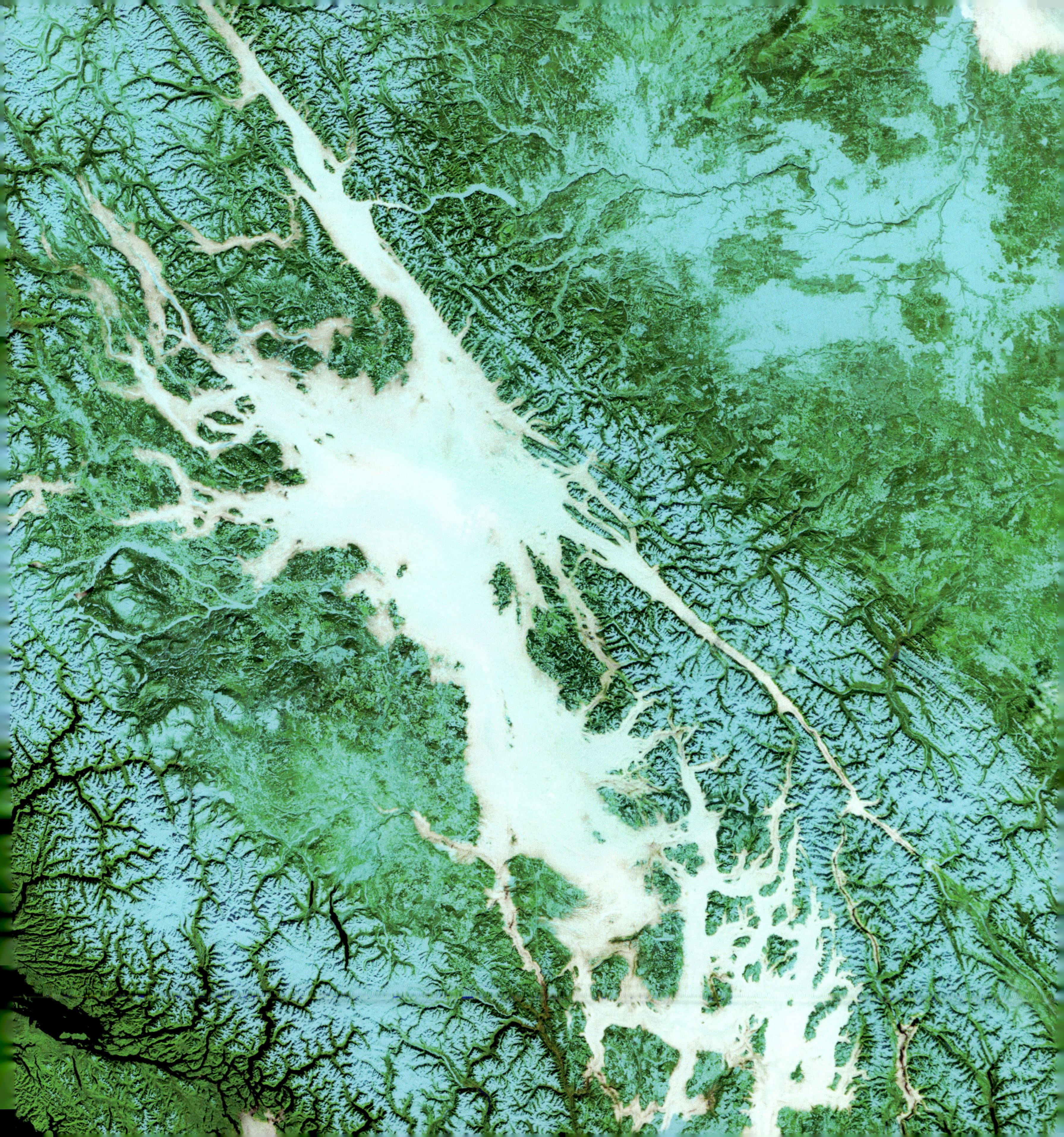

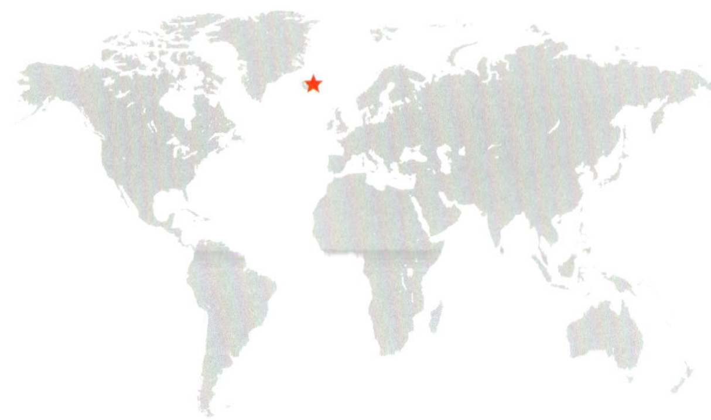

Holuhraun Lava Field
Iceland

As an island in the moist, turbulent North Atlantic, Iceland is often shrouded in clouds and difficult to observe from space. In 2014, the island started making some of its own cloud cover, as the Earth split open between the Bárðarbunga and Askja volcanoes and spewed lava and hot gas.

Landsat 8 captured this view of the eruption in September 2014. The false-color image combines shortwave infrared, near-infrared, and green light. Ice and the plume of steam and sulfur dioxide appear cyan and bright blue, while liquid water is navy blue. Bare or rocky ground around the Holuhraun lava field appears in shades of green or brown, and fresh lava is bright orange. Offshore clouds appear in bright cyan.

Infrared imagery can help scientists estimate the rate at which lava is pouring out of Earth, as well as the sulfur dioxide content of the plume. It also helps them pinpoint lava flows and model how the eruption evolved.

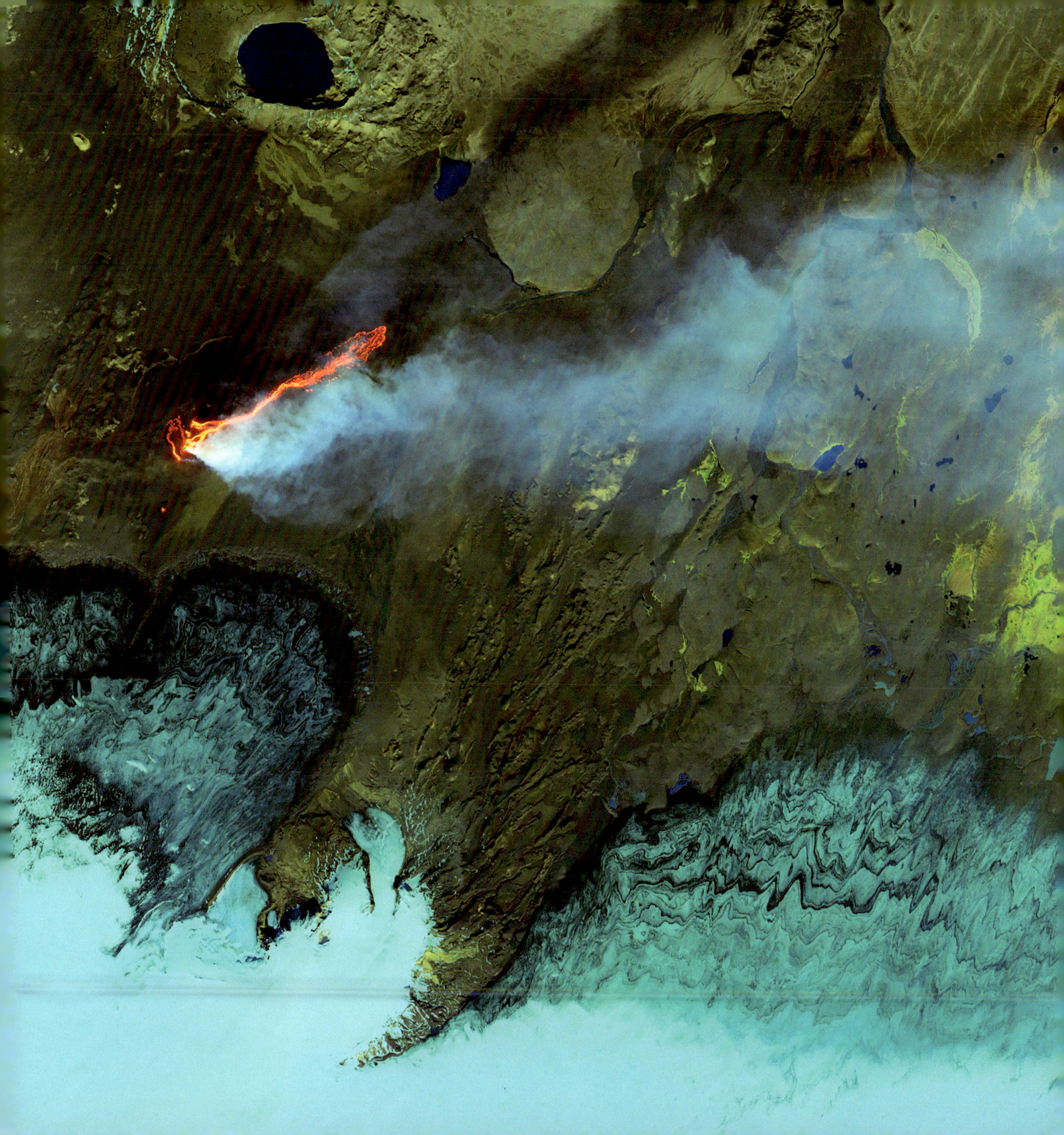

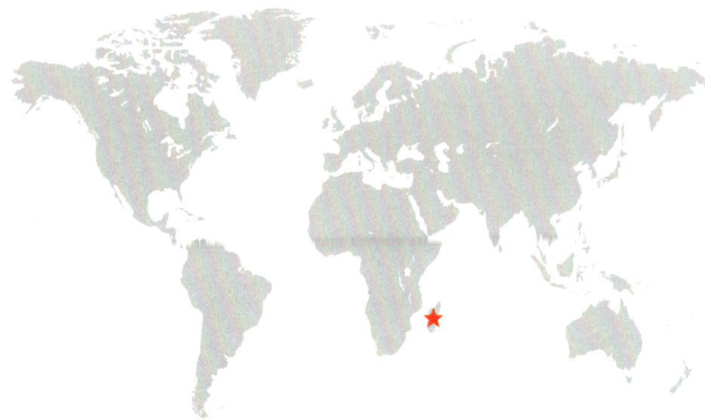

Lofted Over Land
Madagascar

Along the muddy Mania River, midday clouds form over the forested land but not the water. In the tropical rainforests of Madagascar, there is ample moisture for cloud formation. Sunlight heats the land all day, warming that moist air and causing it to rise high into the atmosphere until it cools and condenses into water droplets. Clouds generally form where air is ascending (over land in this case), but not where it is descending (over the river). Landsat 8 acquired this image in January 2015.

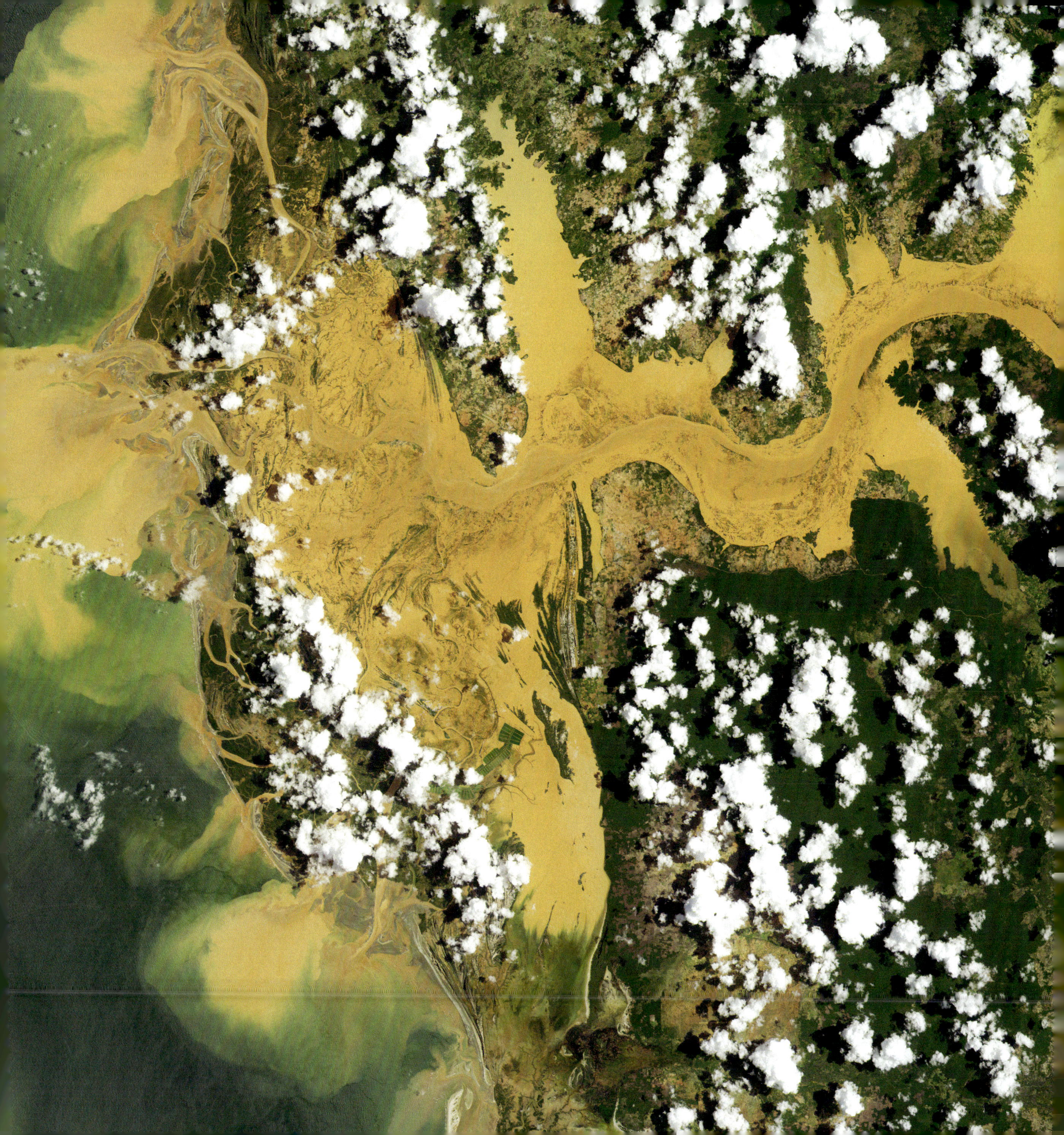

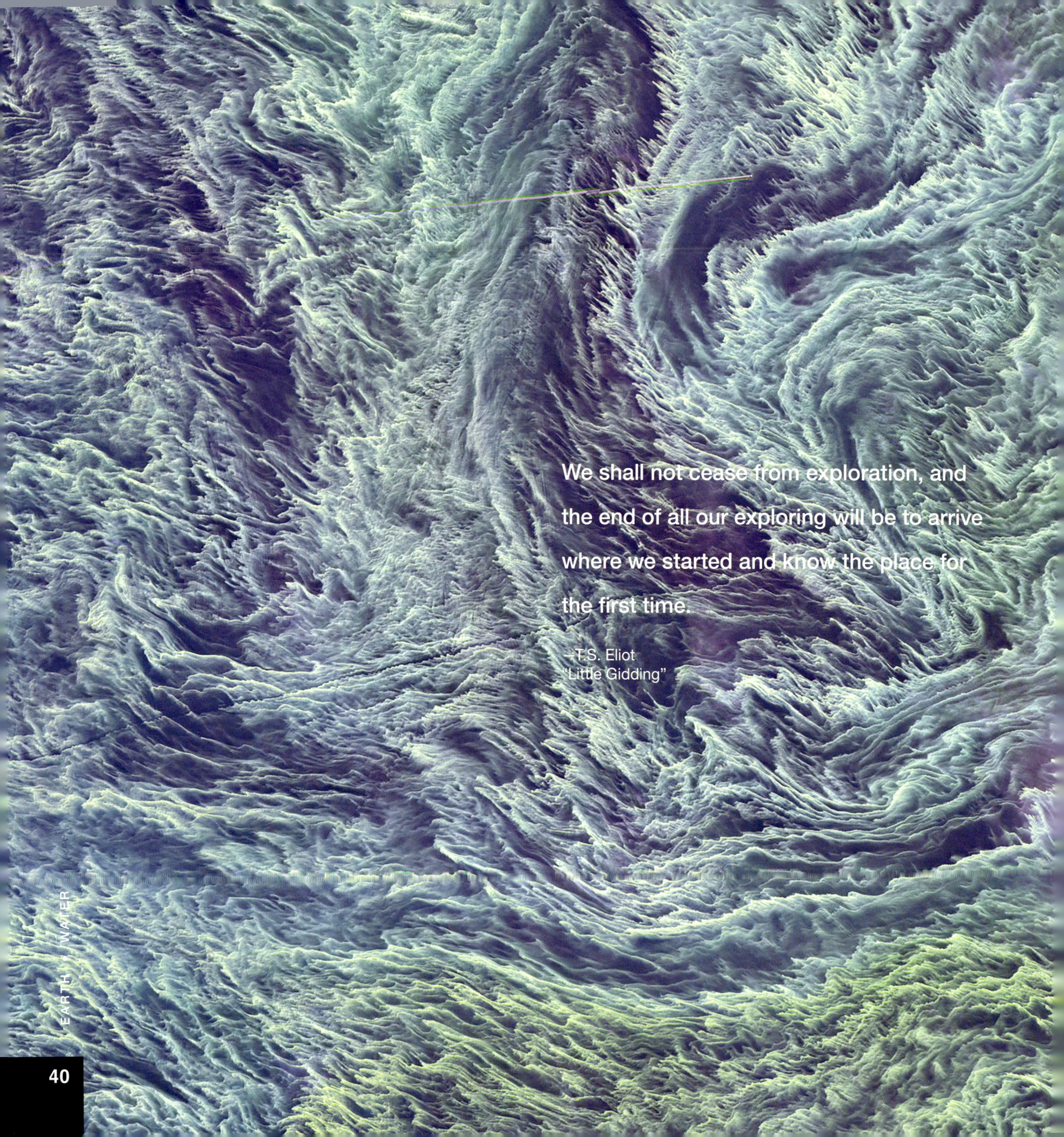

We shall not cease from exploration, and the end of all our exploring will be to arrive where we started and know the place for the first time.

—T.S. Eliot
"Little Gidding"

water

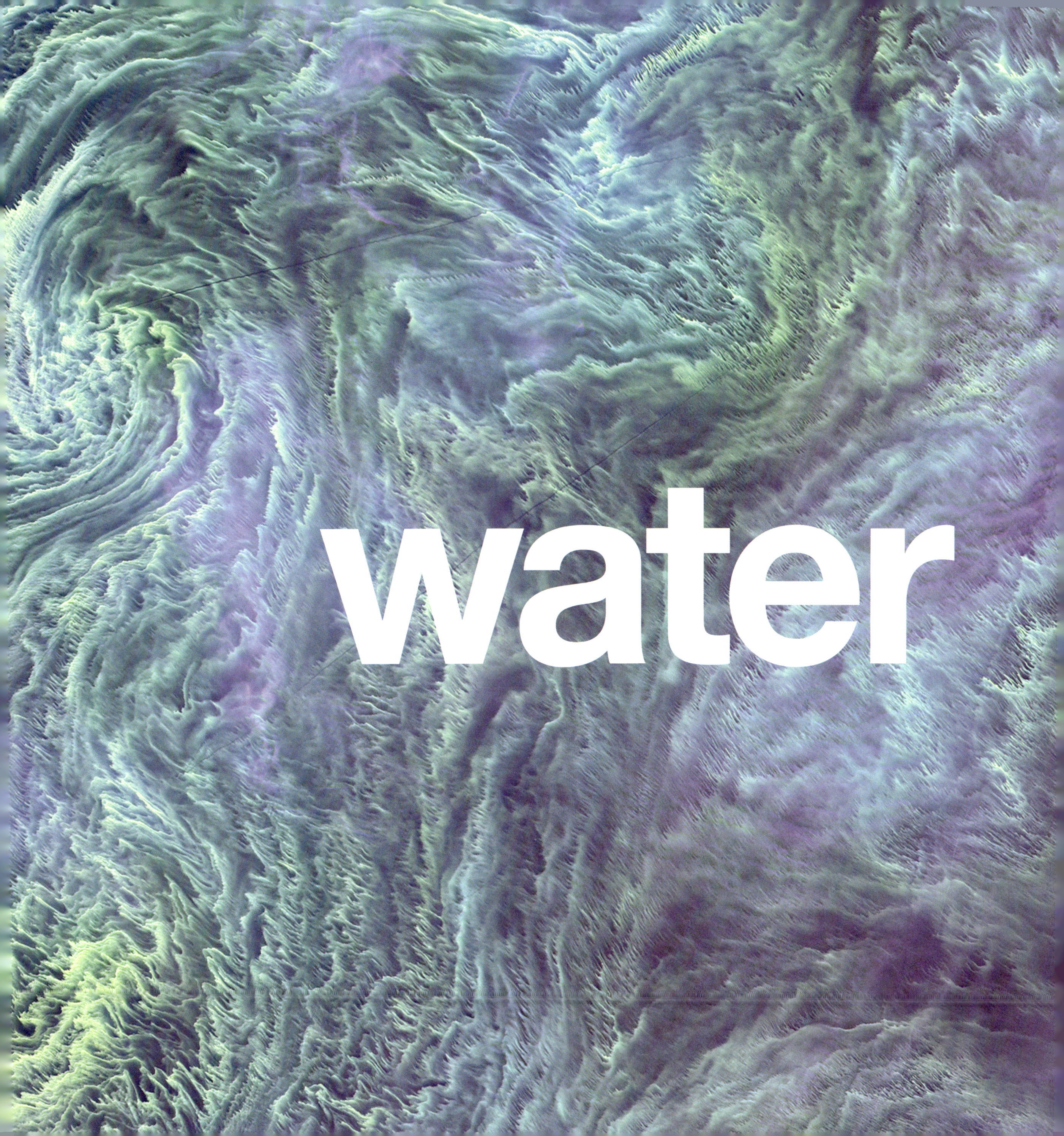

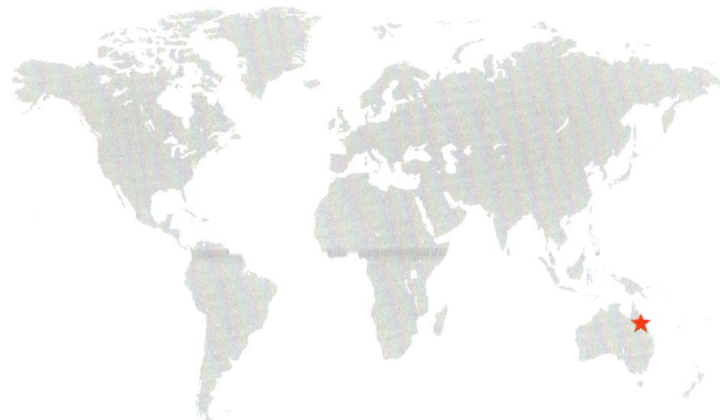

Channel Country
Australia

Australia's Channel Country is full of hundreds of channels, as the Georgina, Burke, and Hamilton rivers merge into the very broad floodplain of Eyre Creek. The land is flat and the drainage is poor, which encourages semi-permanent wetlands to form at the meeting points of the rivers.

These wide floodplains in Queensland are unique on the planet. Scientists think they are caused by the extreme variation in water and sediment discharges from the rivers. In many years there is no rainfall at all, and the rivers are effectively non-existent. In years of modest rainfall, the main channels will carry some water, sometimes spilling over into narrow water holes known as billabongs.

Every few decades, the floodplain carries extremely high discharges of water. For instance, tropical storms to the north can lead to great water flows that inundate the entire width of the floodplain. On such occasions, the floodplain appears as series of brown and green water surfaces with only tree tops indicating the location of the islands. Such is the case in this image taken from the International Space Station in September 2016.

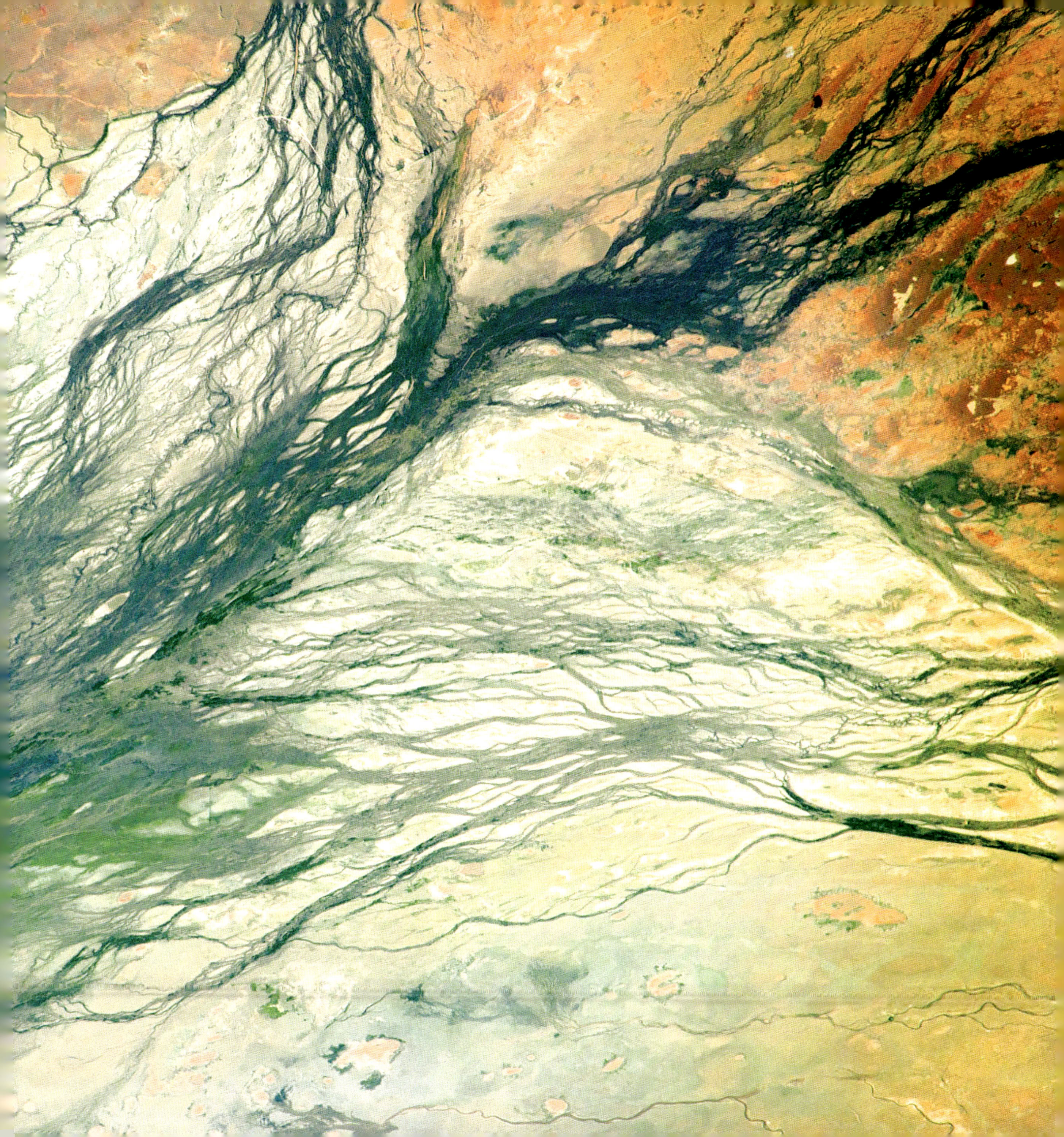

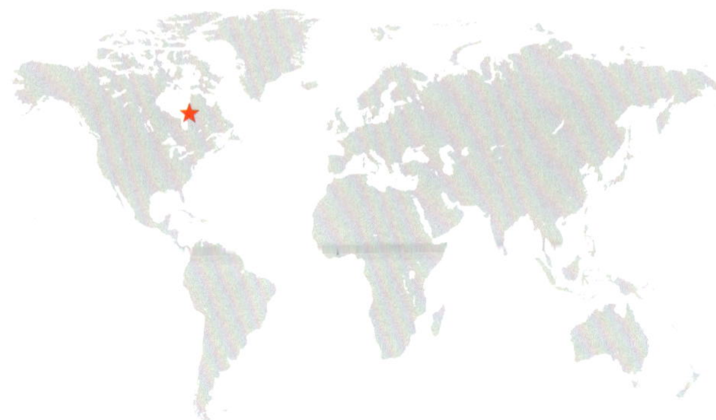

Tea-Colored Rupert Bay
Canada

Remote Rupert Bay is a place where the majesty and dynamism of fluid dynamics is regularly on display. With several rivers pouring into this nook of James Bay, the collision of river and sea water combines with the churn of tides and the motion of currents to make swirls of colorful fluid.

As they wind through the boreal forests and wetlands of northern Quebec, the rivers that flow into Rupert Bay carry water stained brown with natural chemical substances found in plants. Tannins and lignins from roots, leaves, seeds, bark, and soil can leach into the water and give it a yellow, brown, or even black color. (The same process gives tea its dark color.) Note that the colored plumes and intricate vortices around the islands are pointing inland—an indicator that the tide was likely coming in, or that northwesterly winds were affecting the flow of the water.

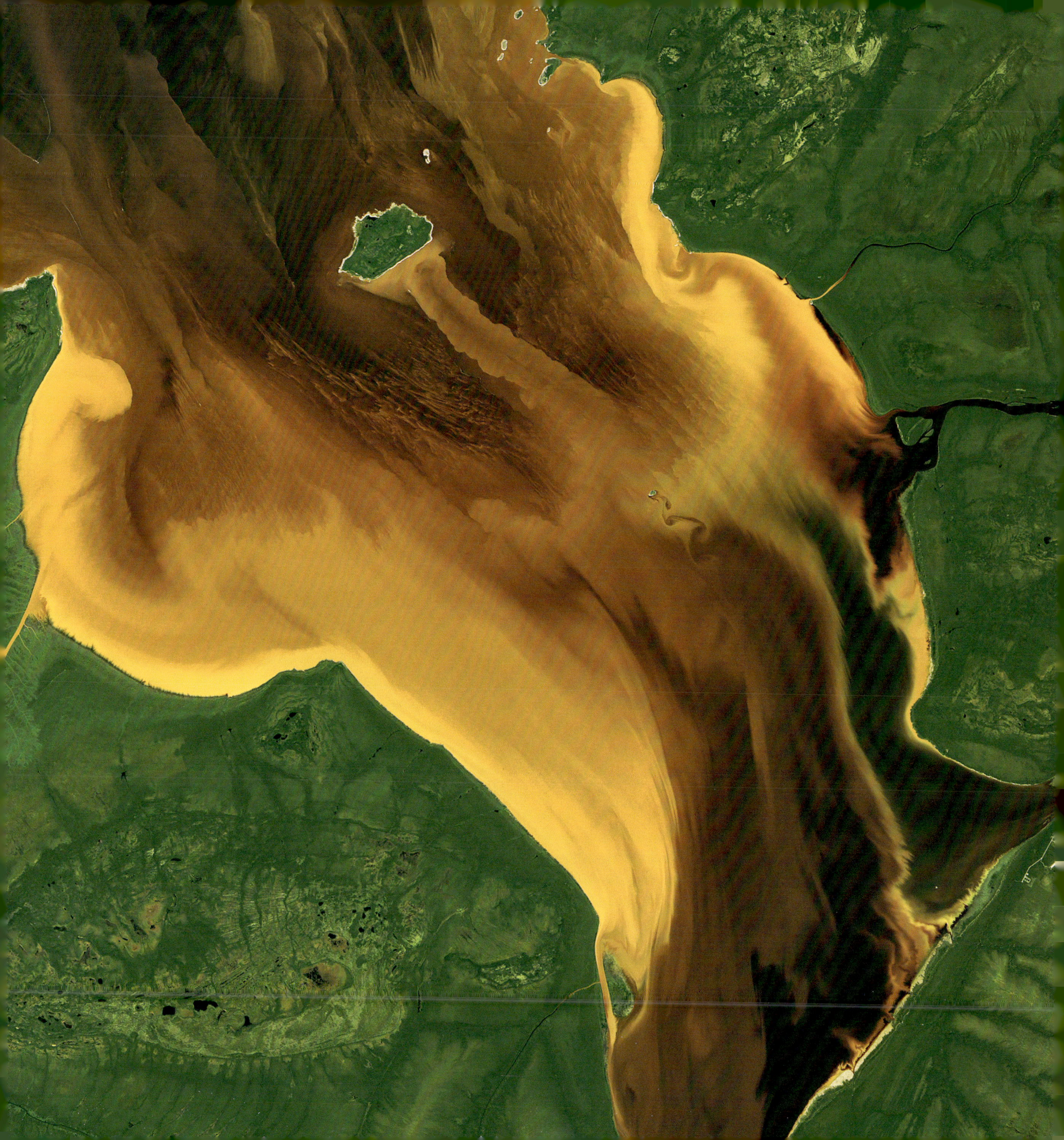

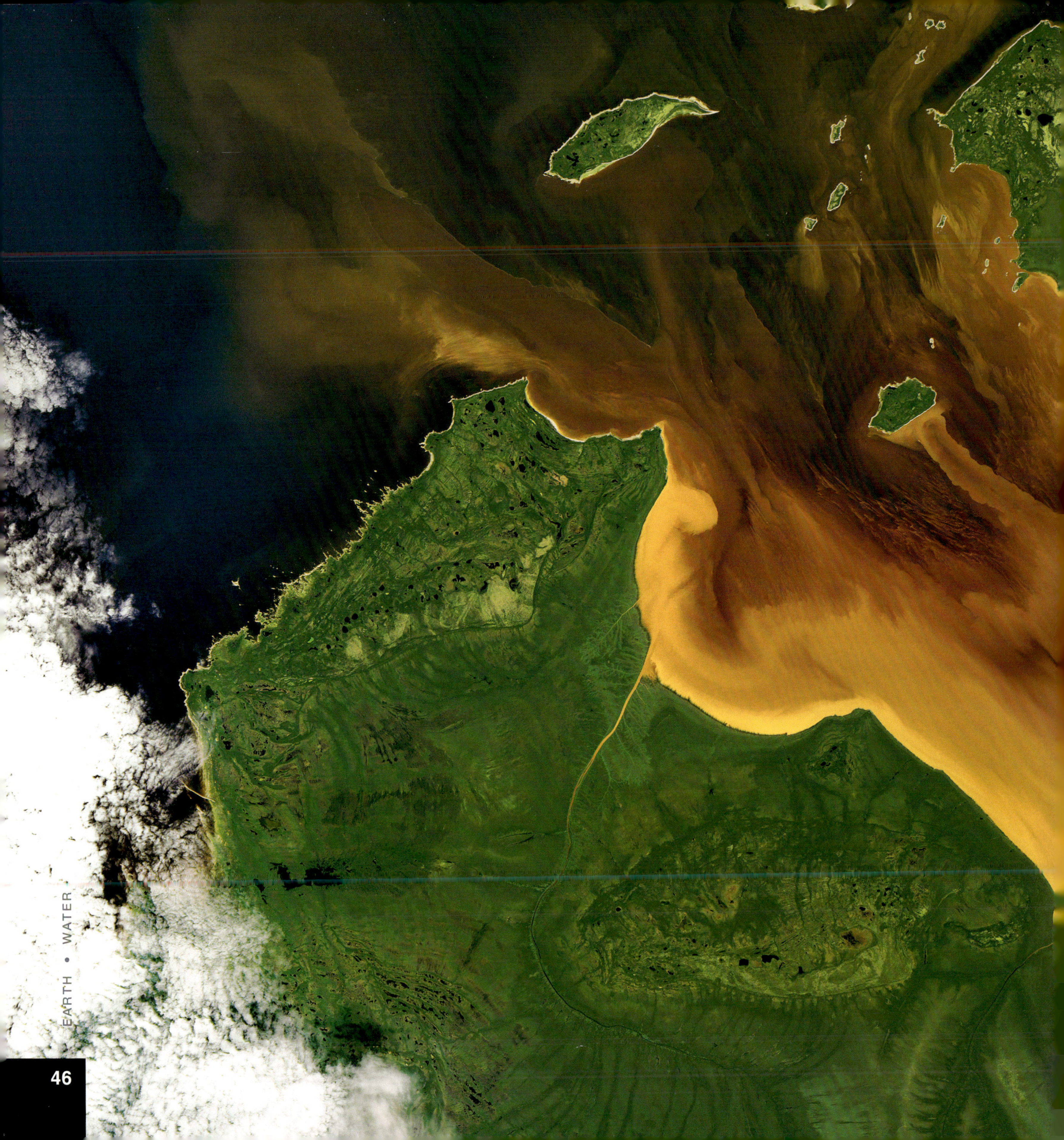

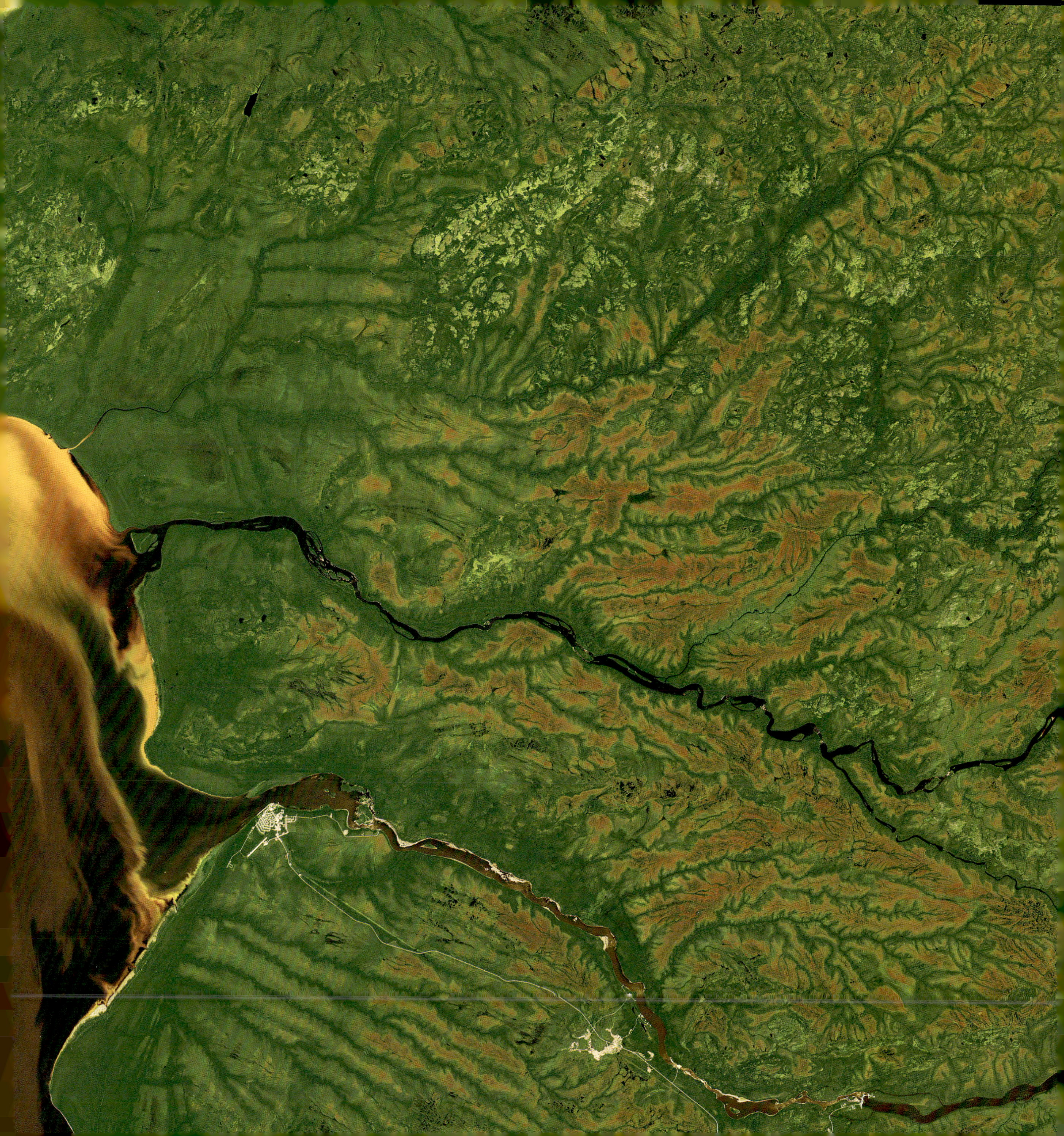

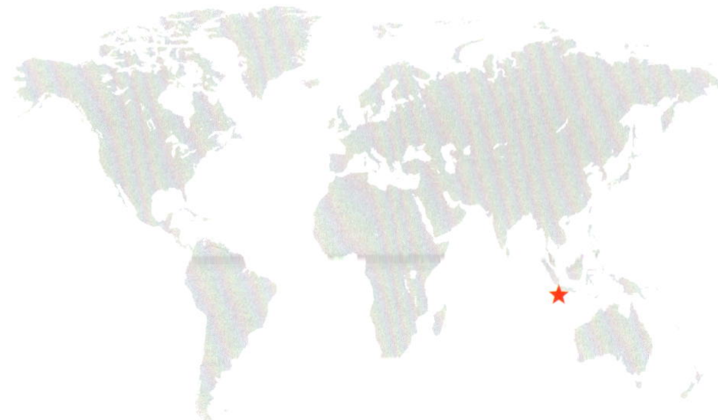

Coral Cocos
Indian Ocean

Coral atolls—which are largely composed of huge colonies of tiny animals such as cnidaria—form around islands. After the islands sink, the coral remains, generally forming complete or partial rings. The South Keeling Islands, part of the Cocos Islands in the Indian Ocean, are such a place.

Only some parts of the South Keeling Islands still stand above the water surface. In the north, the ocean overtops the coral. Along the southern rim of this atoll, shallow water appears aquamarine. Water darkens to navy blue as it deepens toward the central lagoon. Above the water line, coconut palms and other plants form a thick carpet of vegetation. Hard and soft corals thrive throughout the reef.

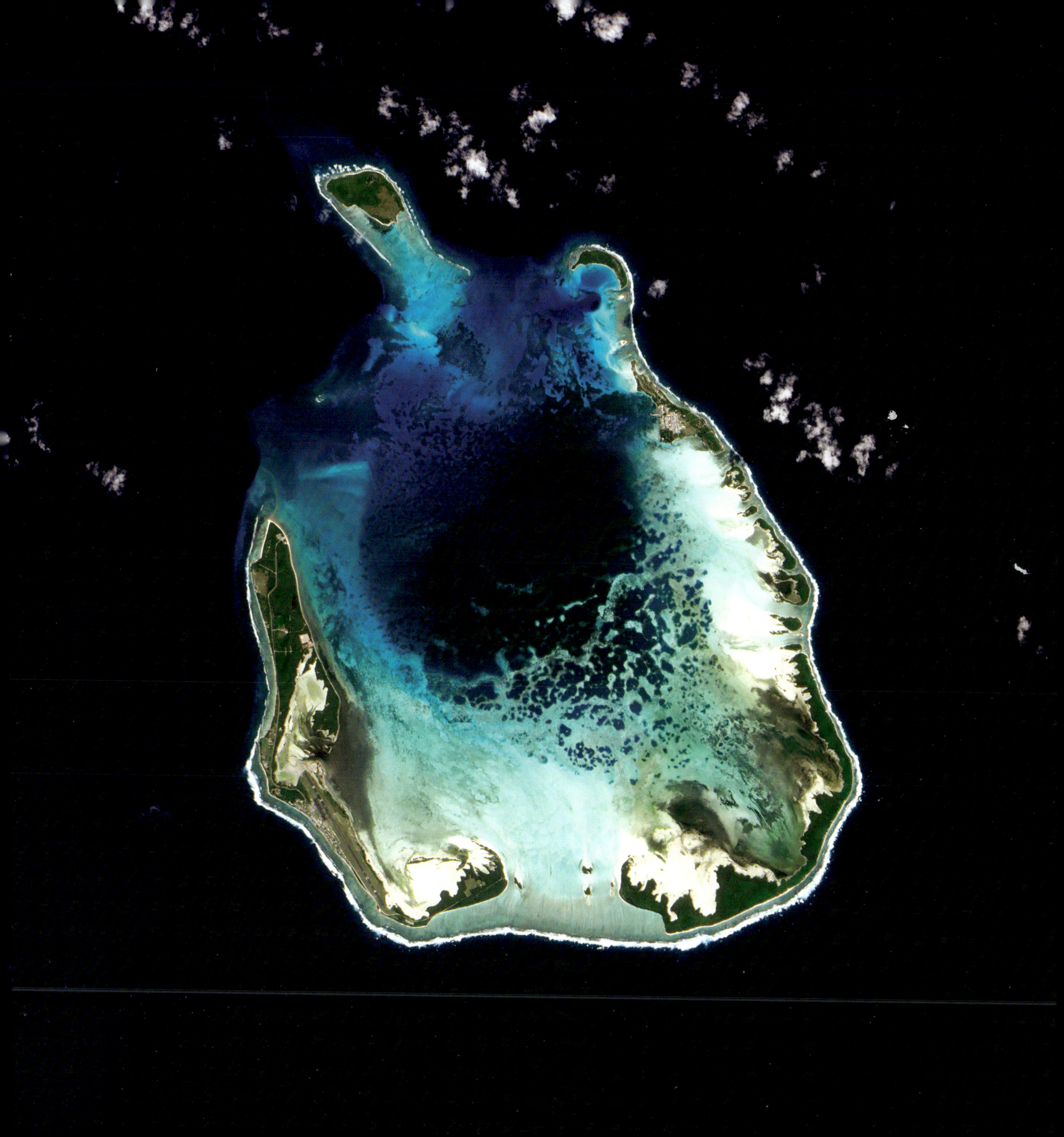

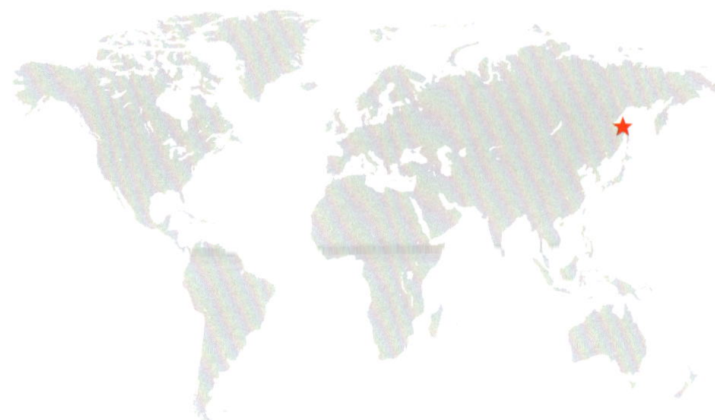

Bay of Whales
Russia

The area around Russia's Ulbanskiy Bay is mostly uninhabited by humans, but it does support sizable numbers of whales. In summertime, this bay is a feeding ground for bowheads, belugas, and orcas that come to Ulbanskiy for the seafood buffet. They hunt by driving fish like herring and smelt toward the coast and into freshwater inlets.

Onshore, freshwater streams meander into marshlands and gently sloped mud flats. The marshes around the bay are dotted with small bodies of water. These are likely thermokarst lakes—pools of water that fill in depressions in the land surface as permafrost melts. Farther inland, the incline becomes steeper and the landscape darker with the greens of pine-covered slopes. A lighter green band (left side) indicates deciduous (leaf-shedding) trees that have begun to turn color.

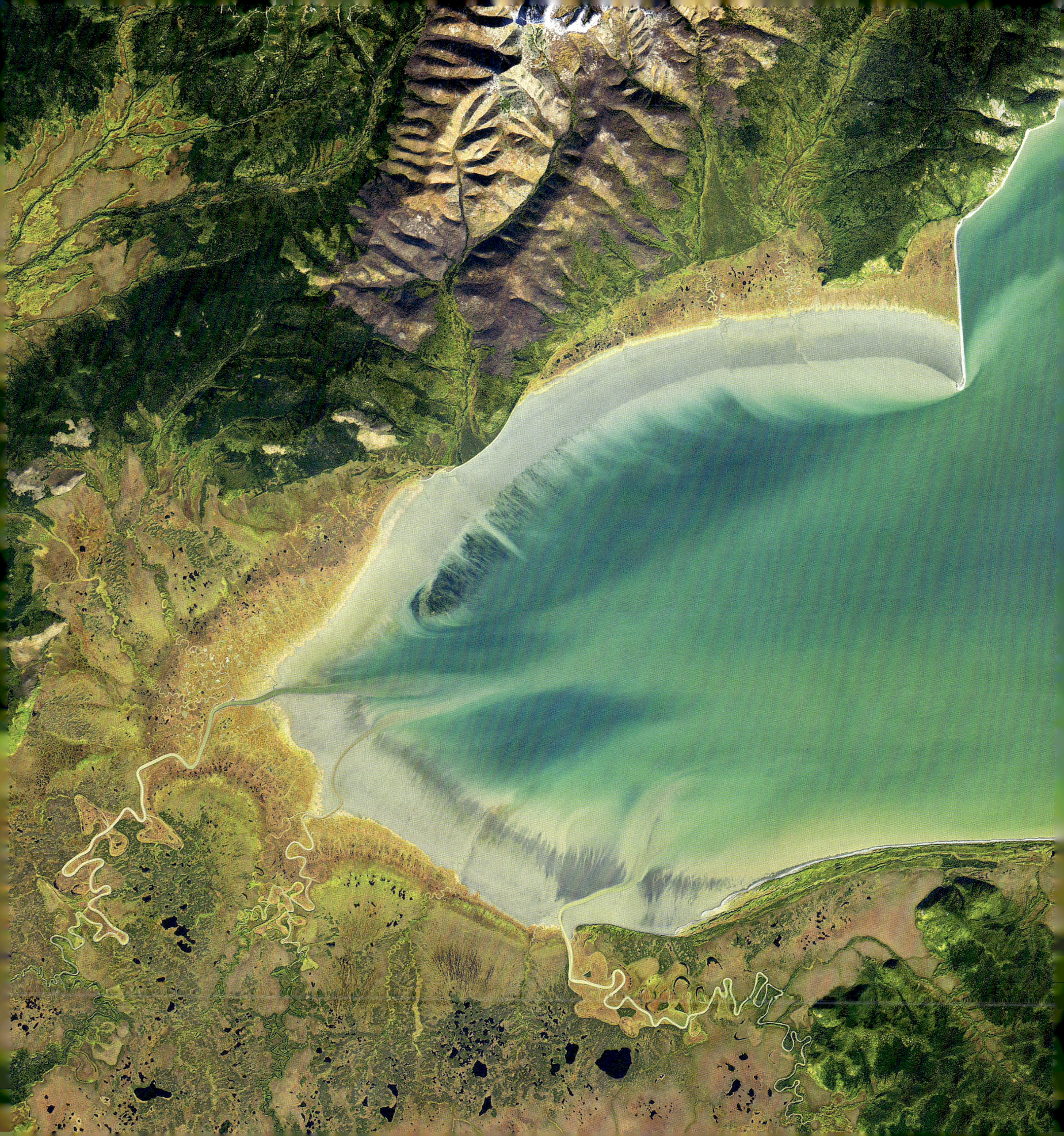

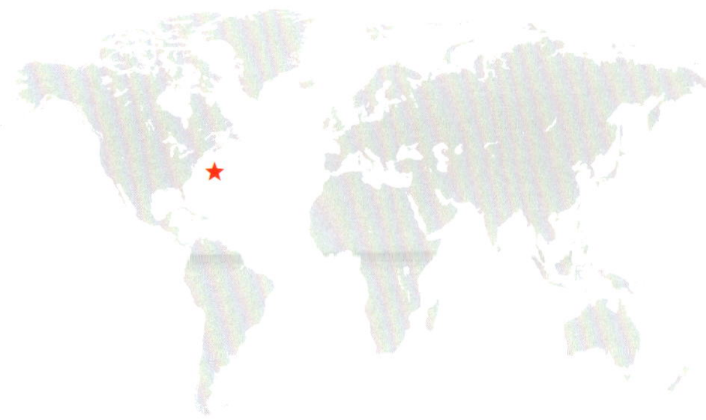

Storms Stir Up Sediment
Bermuda

In October 2014, the eye of Hurricane Gonzalo passed right over Bermuda. In the process, the potent storm stirred up the sediments in the shallow bays and lagoons around the island, spreading a huge mass of sediment across the North Atlantic Ocean. This Landsat 8 image shows the area after Gonzalo passed through.

The suspended sediments were likely a combination of beach sand and carbonate sediments from around the shallows and reefs. Coral reefs can produce large amounts of calcium carbonate, which stays on the reef flats (where there are coralline algae that also produce carbonate) and builds up over time to form islands.

Storm-induced export of carbonates into the deep ocean—where they mostly dissolve—is one of the ways that the oceans naturally balance the addition of atmospheric carbon dioxide to ocean waters.

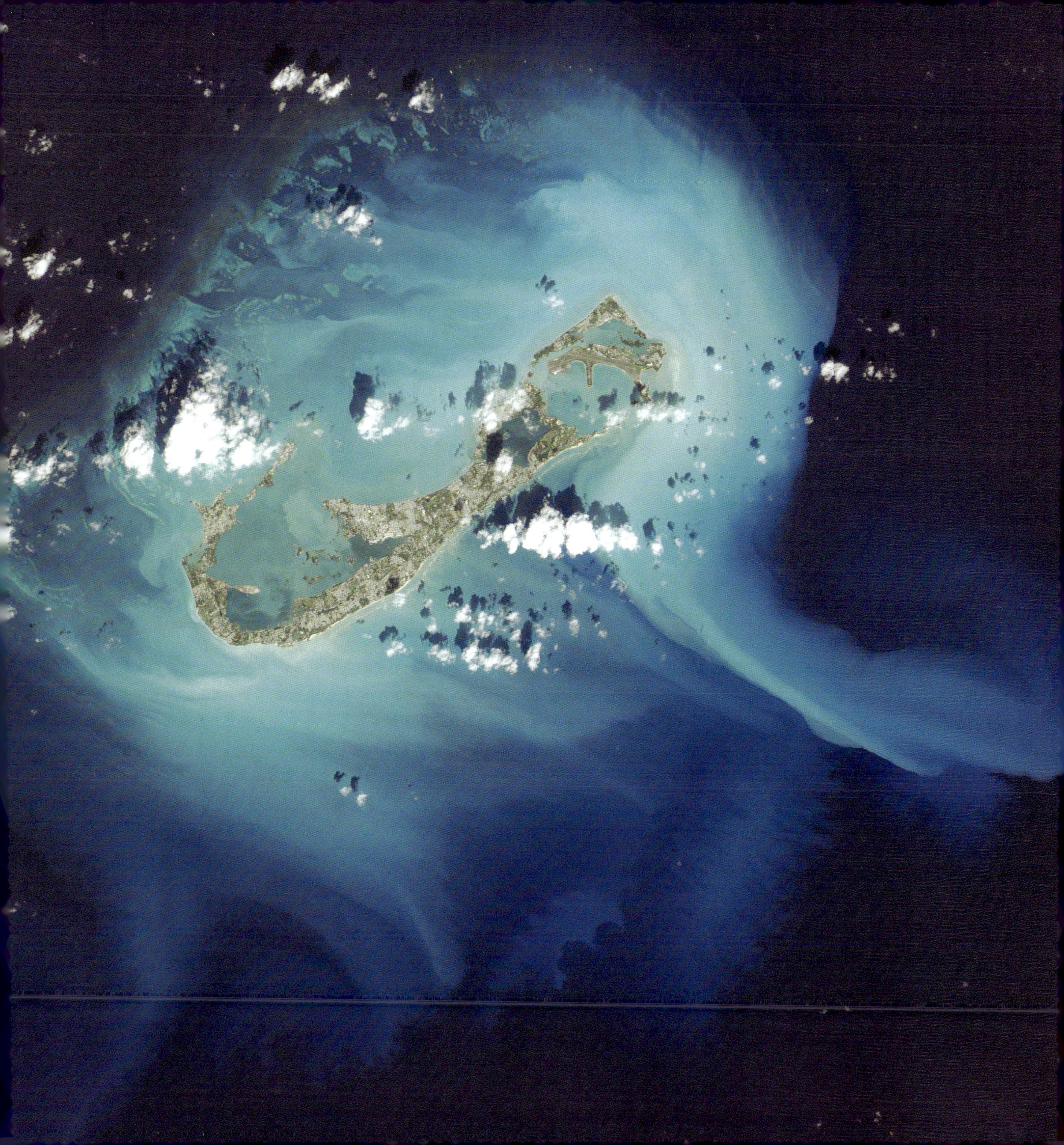

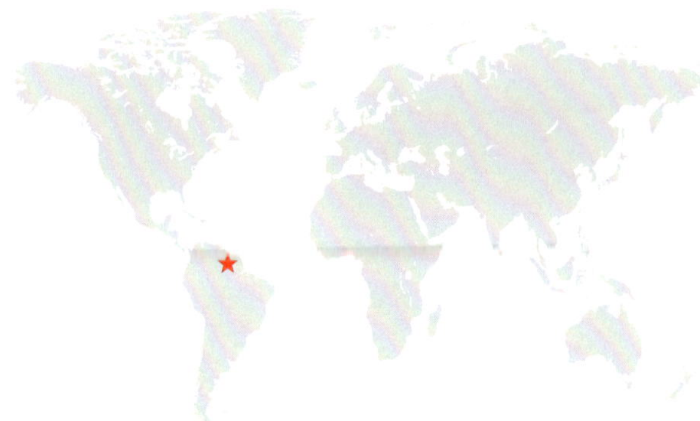

The Meeting of the Waters
Brazil

The Encontro das Águas, or the "Meeting of the Waters," is one of those places that simply wow people.

As Robert Meade of the U.S. Geological Survey once described: "Six Mississippi Rivers' worth of cafe-au-lait-colored water are converging here with two Mississippis' worth of black-tea-colored water to produce the greatest hydrologic spectacle on the planet."

The coffee-colored Rio Solimões, rich with sediment, runs down from the Andes Mountains. The black-tea-tinted Rio Negro that flows from the Colombian hills and jungles is nearly sediment-free and colored by decayed leaf and plant matter. Where the rivers meet, east of Manaus, Brazil, they flow side by side within the same channel for several kilometers. The cooler, denser, and faster waters of the Solimões and the warmer, slower waters of the Negro form a boundary that is visible from space. Turbulent eddies eventually mix the two and become the Lower Amazon River.

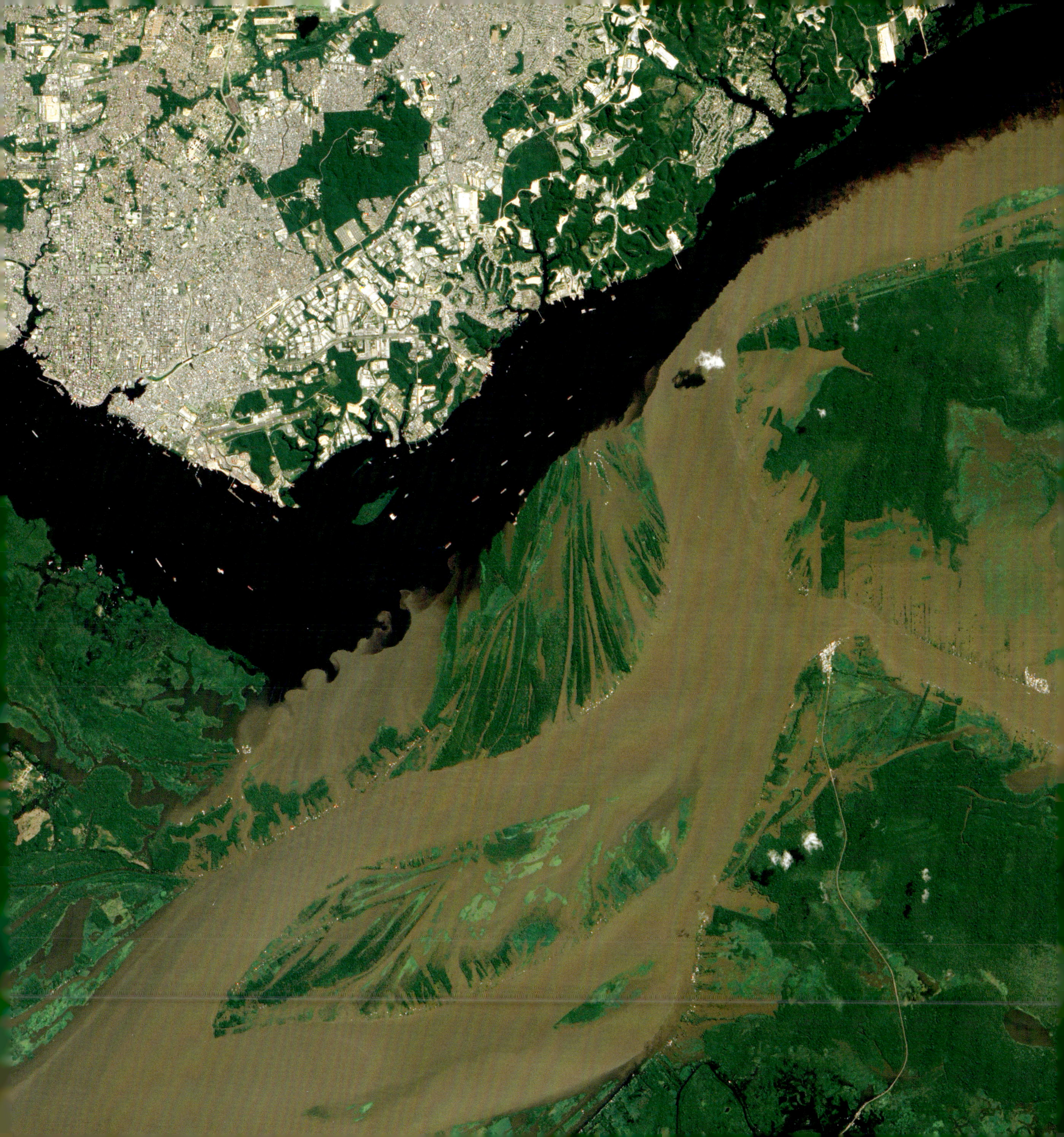

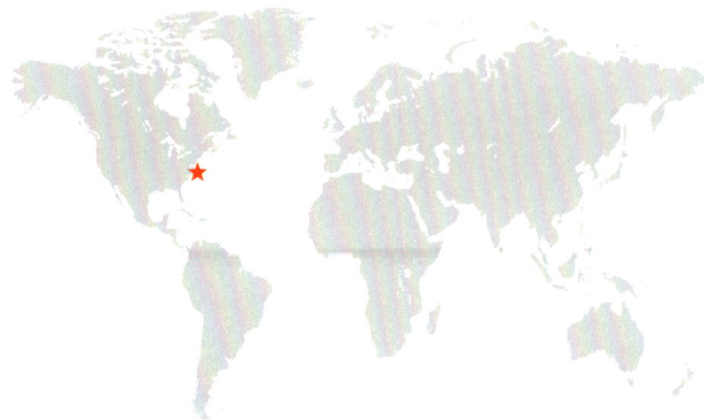

A Lava Lamp Look at the Atlantic
Atlantic Ocean

Stretching from tropical Florida to the doorstep of Europe, the Gulf Stream carries a lot of heat, salt, and history. This river of water is an important part of the global ocean conveyor belt, moving water and heat from the Equator toward the far North Atlantic. It is one of the strongest currents on Earth and one of the most studied. Its discovery is often attributed to Benjamin Franklin, though sailors likely knew about the current long before they had a name for it.

This image shows a small portion of the Gulf Stream off of South Carolina as it appeared in infrared data collected by the Landsat 8 satellite in April 2013. Colors represent the energy—heat—being emitted by the water, with cooler temperatures in purple and the warmest water being nearly white. Note how the Gulf Stream is not a uniform band but instead has finer streams and pockets of warmer and colder water.

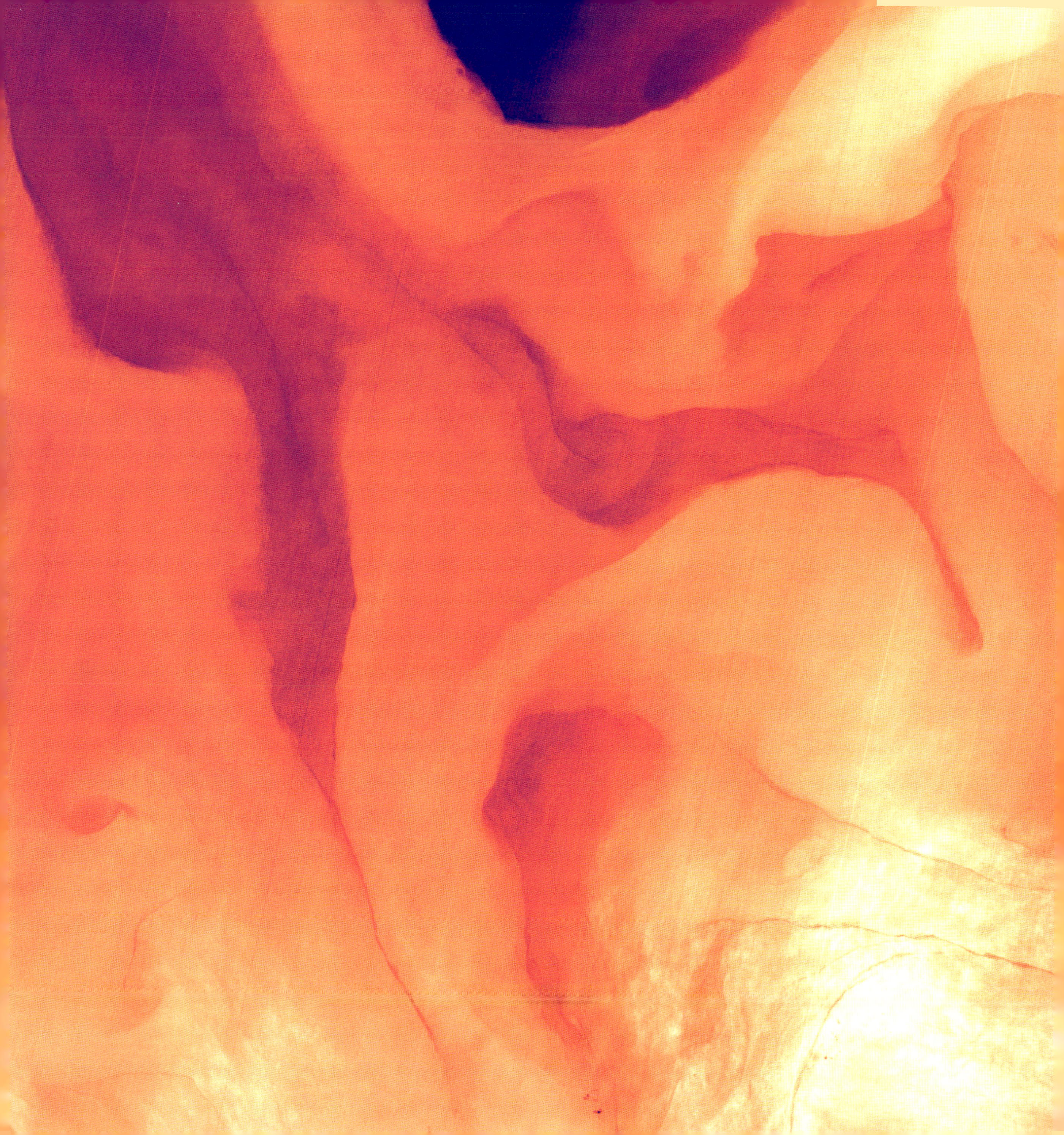

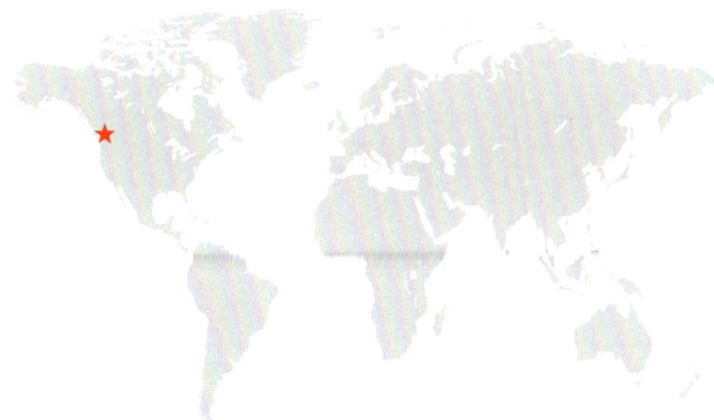

Teeming Life in the Strait of Georgia
Canada

In August 2016, the waters off of British Columbia turned bright green. The Strait of Georgia and nearby inlets were teeming with coccolithophores—a harmless type of floating plant-like organisms called phytoplankton. Coccolithophores have chalky, scale-like shells made of calcium carbonate. The milky-white color of those shells can brighten and discolor otherwise blue waters when the plankton explode in such massive blooms.

While coccolithophore blooms occasionally occur off the west coast of Vancouver Island, few scientists could recall seeing a bloom like this in the strait. Research suggests that coccolithophore numbers have been increasing in recent decades even as the water has been growing more acidic.

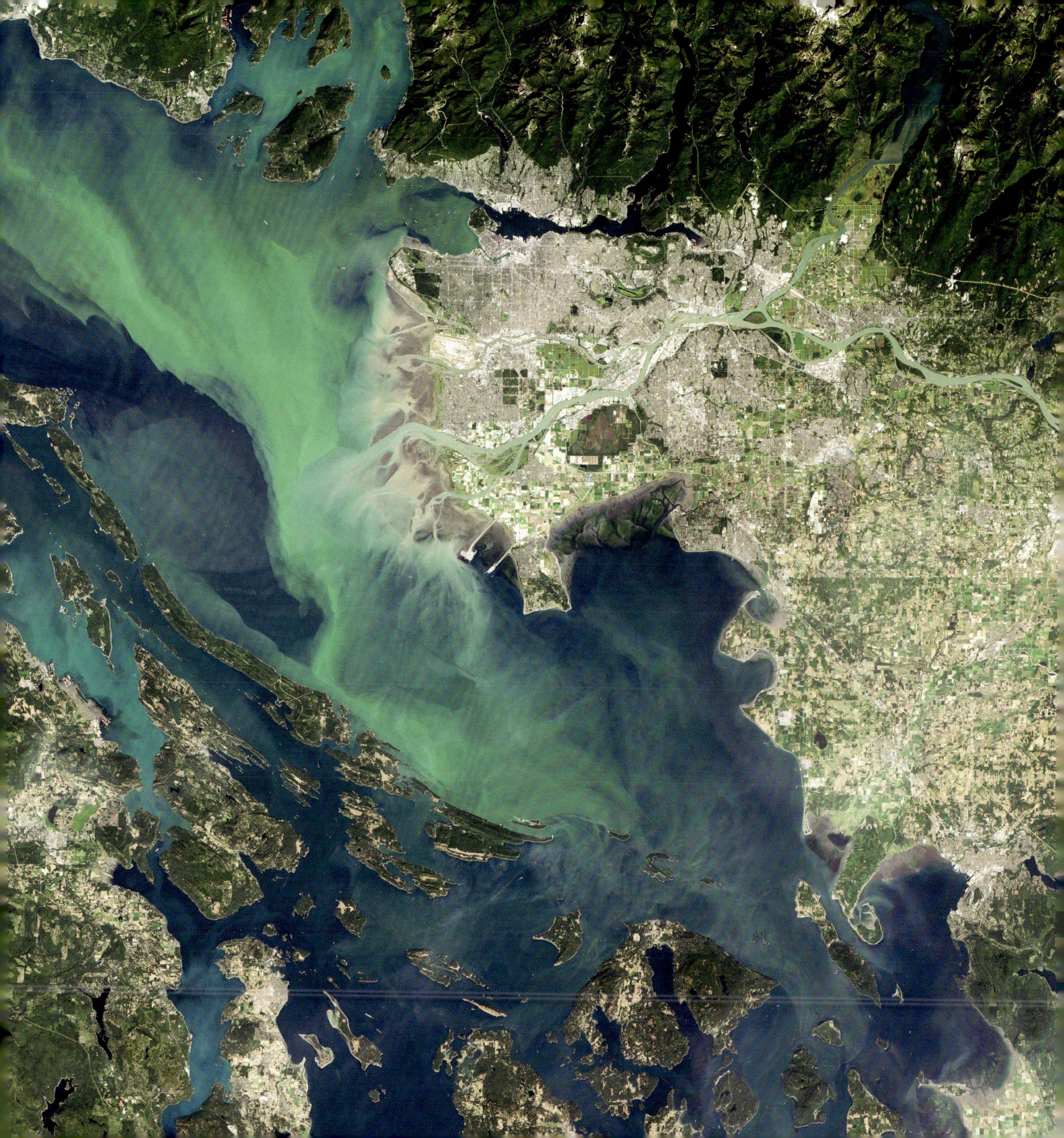

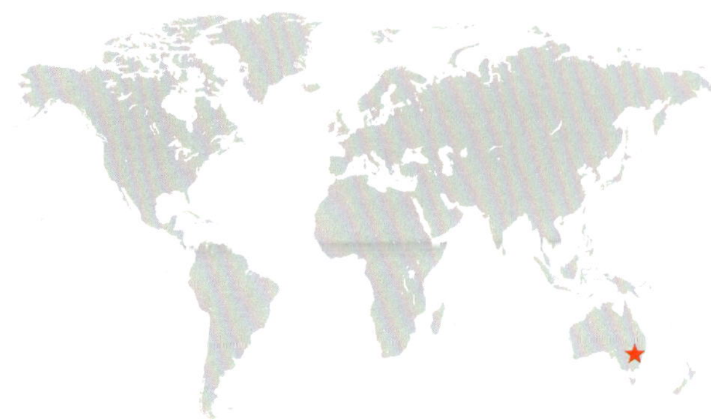

Ephemeral Lake Frome
Australia

The interior of Australia is full of ephemeral lakes. These basins pass most of their time as salt pans, but occasional heavy rains can fill them with water.

The Earth Observing-1 satellite captured this image in April 2010 after water had flowed into Lake Frome, which stands at the southern end of an arc of salt pans. When it fills, the waters usually come from precipitation in the hills and other salt pans upstream. The land to the east has a slightly higher elevation and consists of a network of dry river channels. Inside the salt pan, the land surface is uneven. Areas shaped like sloppy teardrops rise above the surrounding plain. Water on the surface appears in shades of dull green. And throughout much of Lake Frome, water makes its presence known not through standing water but through mud or wet salts. Darker surfaces show where water has seeped through typically dry sediments.

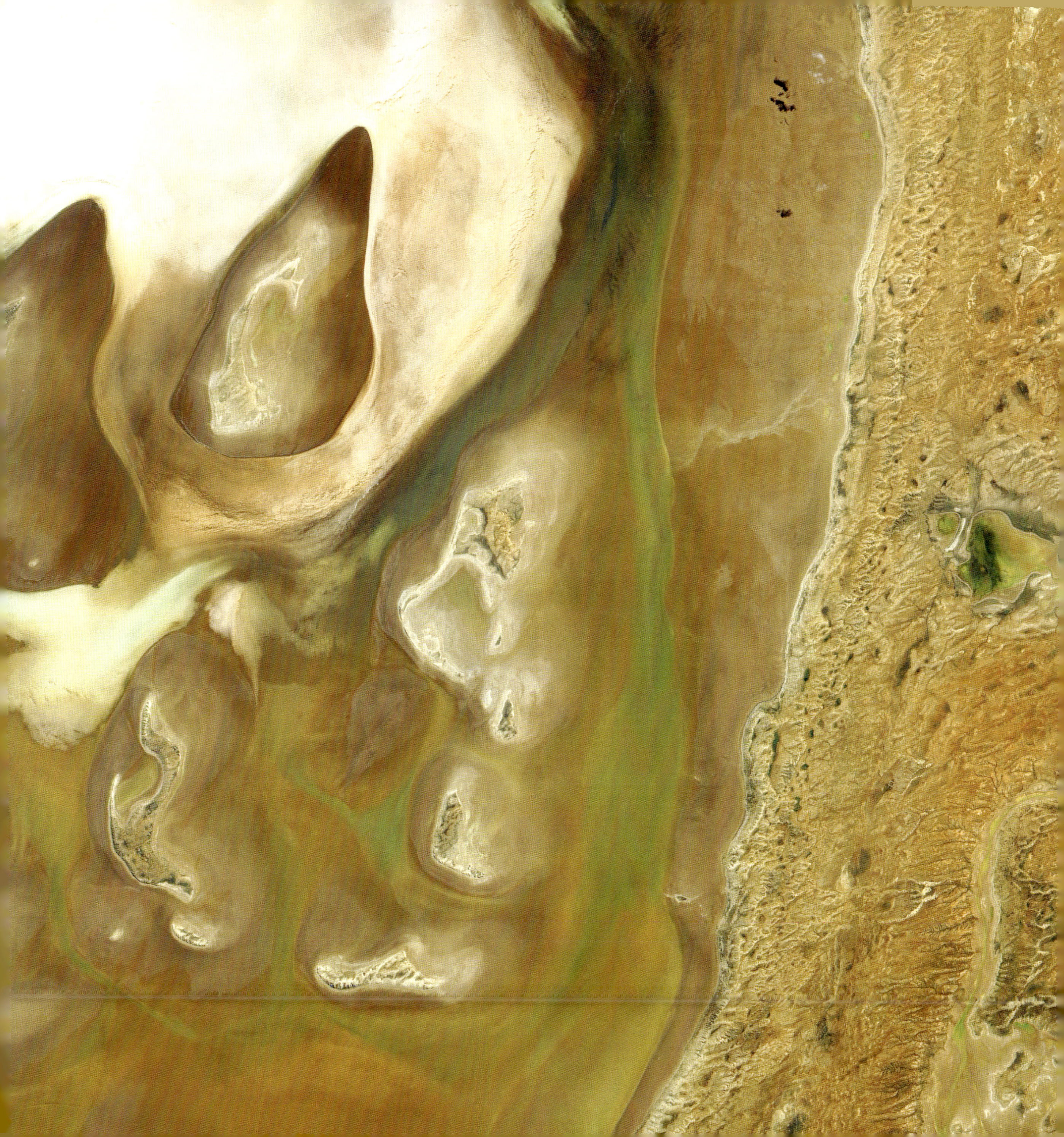

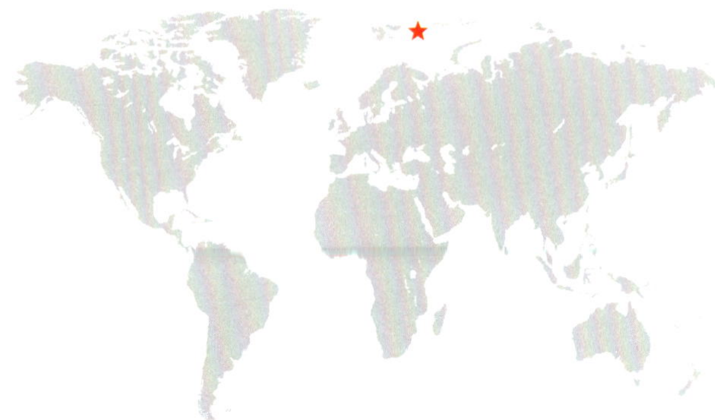

Dueling Blooms
Barents Sea

As the seasons pass on Earth, different species tend to dominate the landscape at different times. Such was the case in July 2014 in the surface waters of the Barents Sea, north of Norway and Russia. The Aqua satellite captured a transitional moment between one form of microscopic, plant-like organisms (phytoplankton) and another.

Several currents merge in this area, and intersecting waters combine with stiff winds to promote mixing of waters and nutrients from the deep. Note the green swirls on the center and left, as well as the milky, blue-white swirls on the upper right. (The fluffy white area is cloud cover.) It is likely that the green plankton were diatoms and the white ones were coccolithophores. Research has suggested that diatoms start to bloom in the well-mixed, cooler waters of spring and dominate the early summer. As the water warms and becomes more stratified or layered, coccolithophores bloom more abundantly.

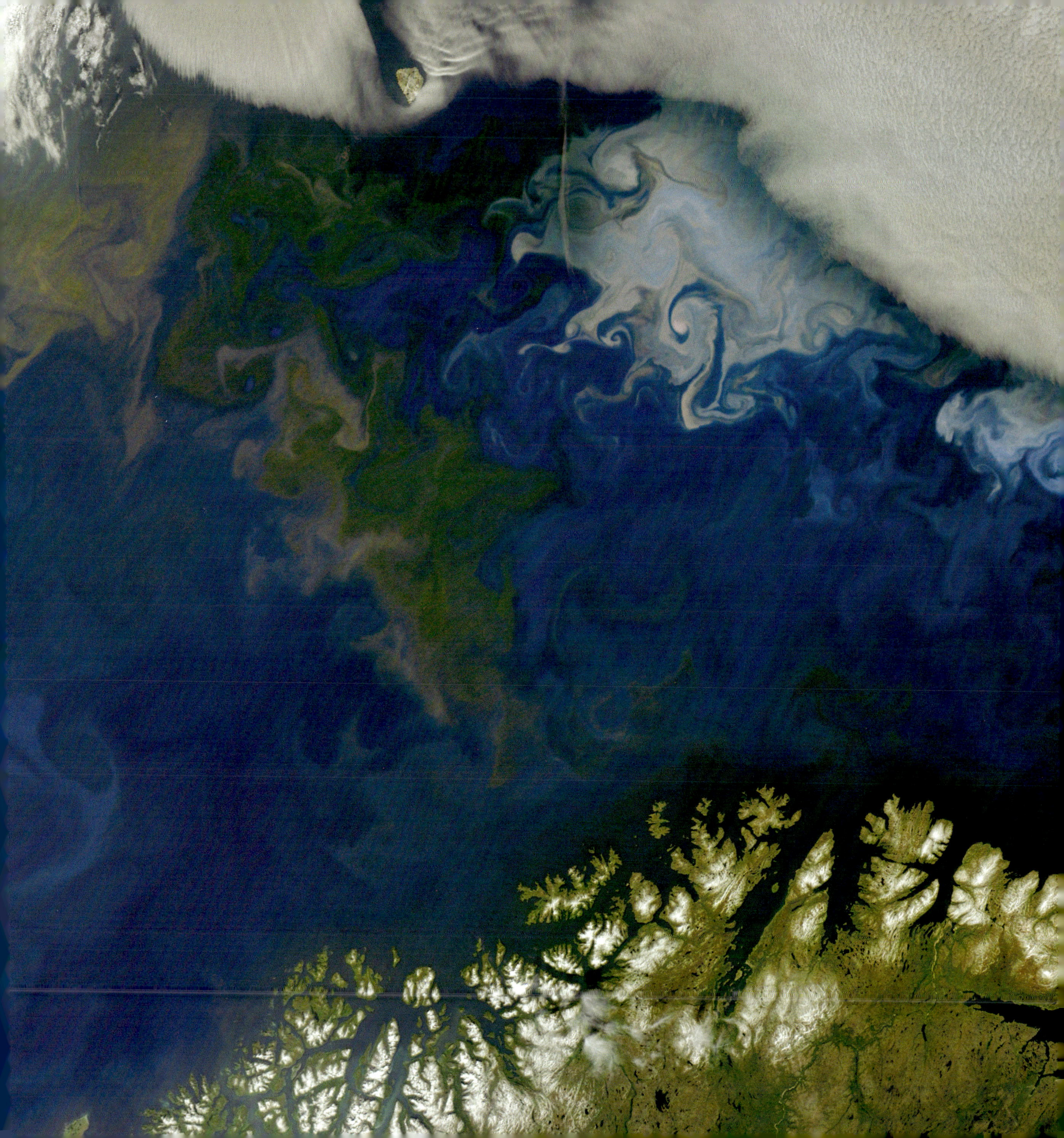

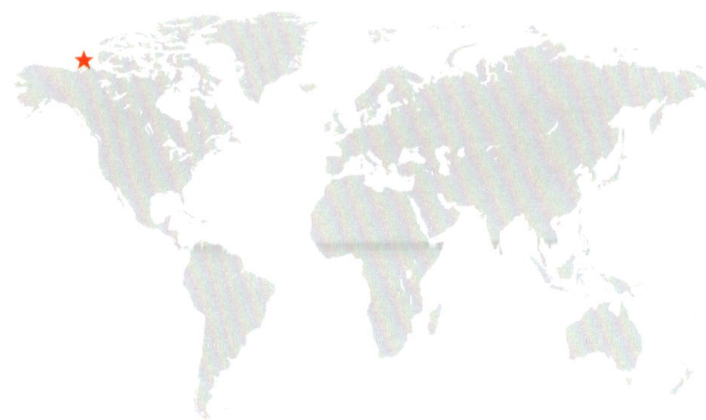

A Bay Sculpted by Ice
Canada

The land around Liverpool Bay in Canada's Northwest Territories owes its otherworldly appearance to ice past and present.

Thousands of years ago, this area was buried under a massive ice sheet that sprawled over much of North America. During that time, glacial activity carved out parallel lakes separated by strips of land that look like giant, skeletal fingers. After the glaciers retreated, pockets of ice lingered underground. As those pockets have melted and the frozen ground has thawed, lakes have formed.

Geologists have offered different explanations for the formation of the finger-shaped ridges: they may be moraines created by the movement of ice, or they might be sediment ridges that formed between channels of subglacial meltwater. Smaller thermokarst lakes in this scene are formed by both the long- and short-term melting of ice and permafrost.

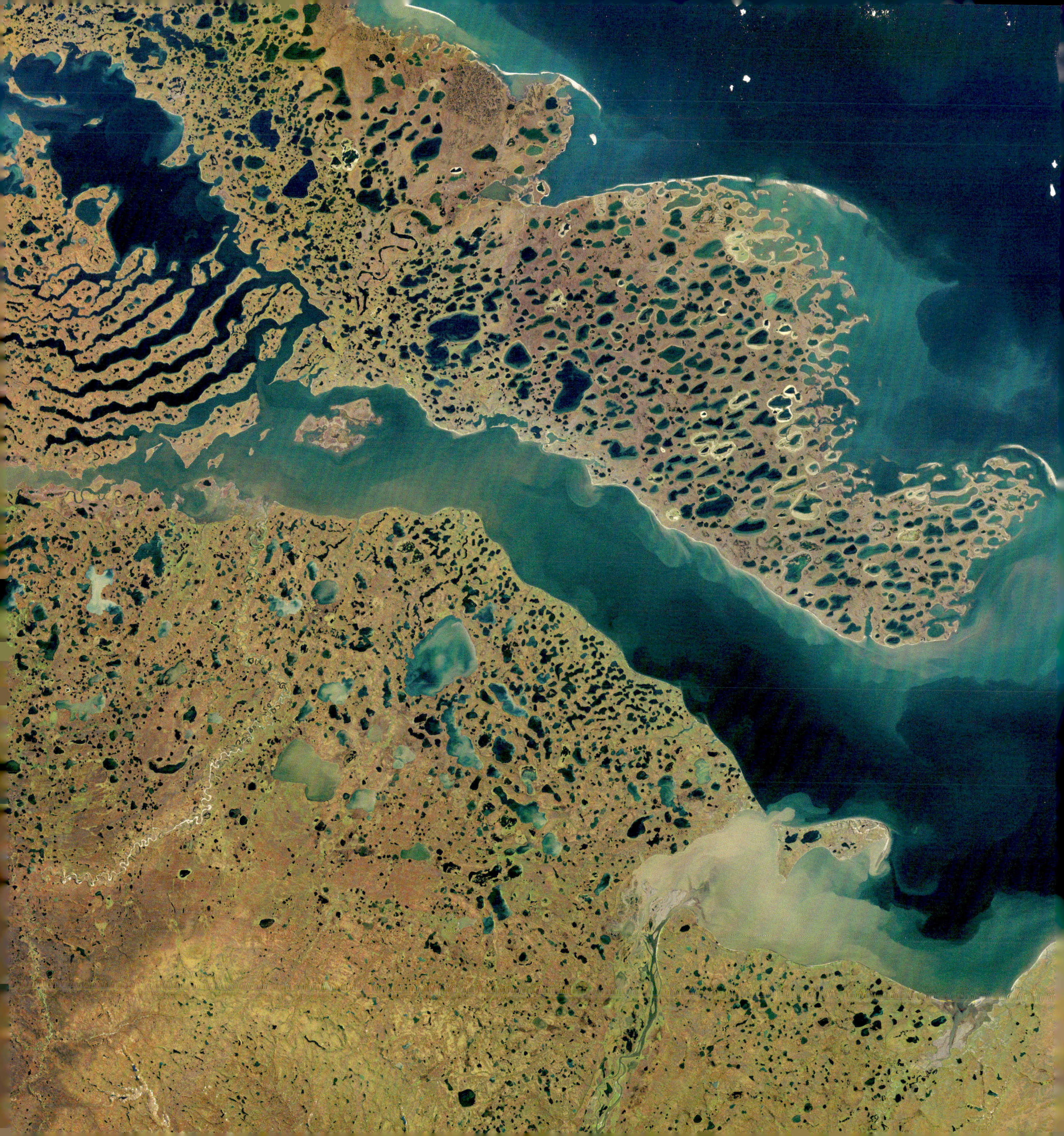

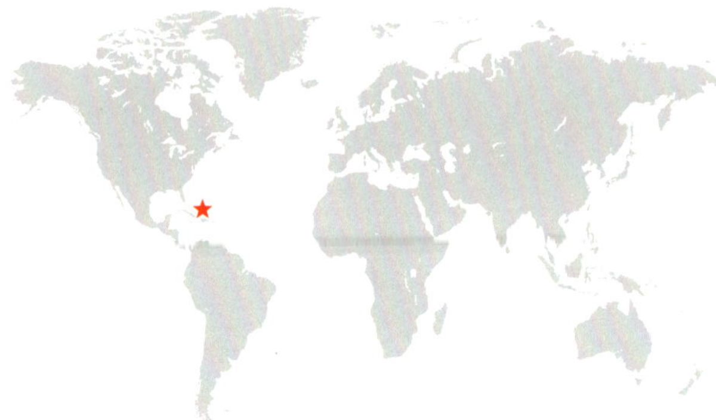

Tidal Flats and Channels
Bahamas

The islands of the Bahamas are situated on large depositional platforms—the Great and Little Bahama Banks—composed mainly of carbonate sediments ringed by reefs. The islands are the only parts of the platform currently exposed above sea level. The sediments were formed mostly from the skeletal remains of organisms settling to the sea floor; over geologic time, these sediments consolidated to form carbonate sedimentary rocks such as limestone.

This November 2010 photograph from the International Space Station provides a view of tidal flats and underwater channels along the eastern margin of the Great Bahama Bank. The continuously exposed parts of the islands are brown, a result of soil formation and vegetation growth. To the north and west, we can see through shallow water to the off-white tidal flats composed of carbonate sediments. The tidal flow of seawater is concentrated through gaps in the seafloor, leading to the formation of relatively deep channels that cut into the sediments.

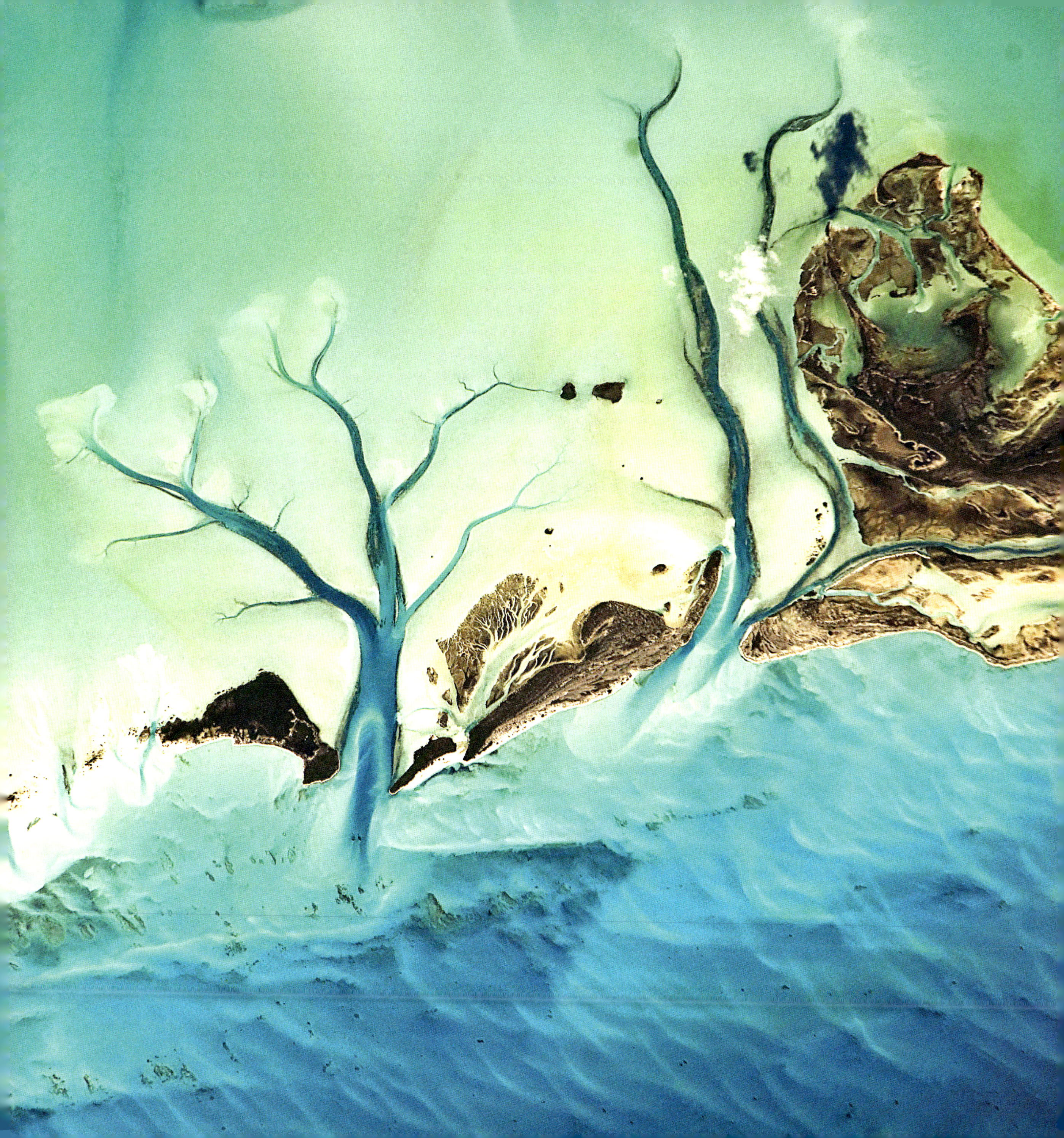

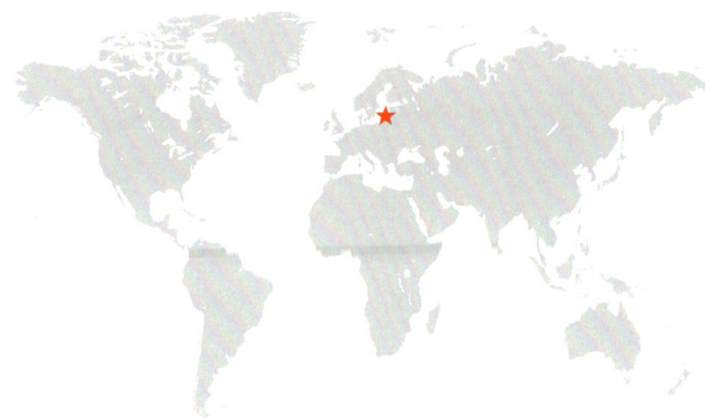

The Blooming Baltic
Baltic Sea

Cyanobacteria are an ancient type of marine bacteria that, like other phytoplankton, capture and store solar energy through photosynthesis. Some cyanobacteria are toxic to humans and animals. Moreover, large blooms can sometimes cause oxygen-depleted dead zones where other organisms cannot survive.

In August 2015, Landsat 8 captured this false-color view of a large bloom of cyanobacteria swirling in the Baltic Sea. Blooms flourish here during summertime, when there is ample sunlight and high levels of nutrients. Tracks of several ships show up as dark lines where they have cut through the bloom.

Agricultural and industrial runoff from Europe can contribute to excess nutrients in the Baltic Sea. Nutrient loads have been decreasing since 1980, and coastal areas have seen improvement, yet concentrations in the open sea have not changed much.

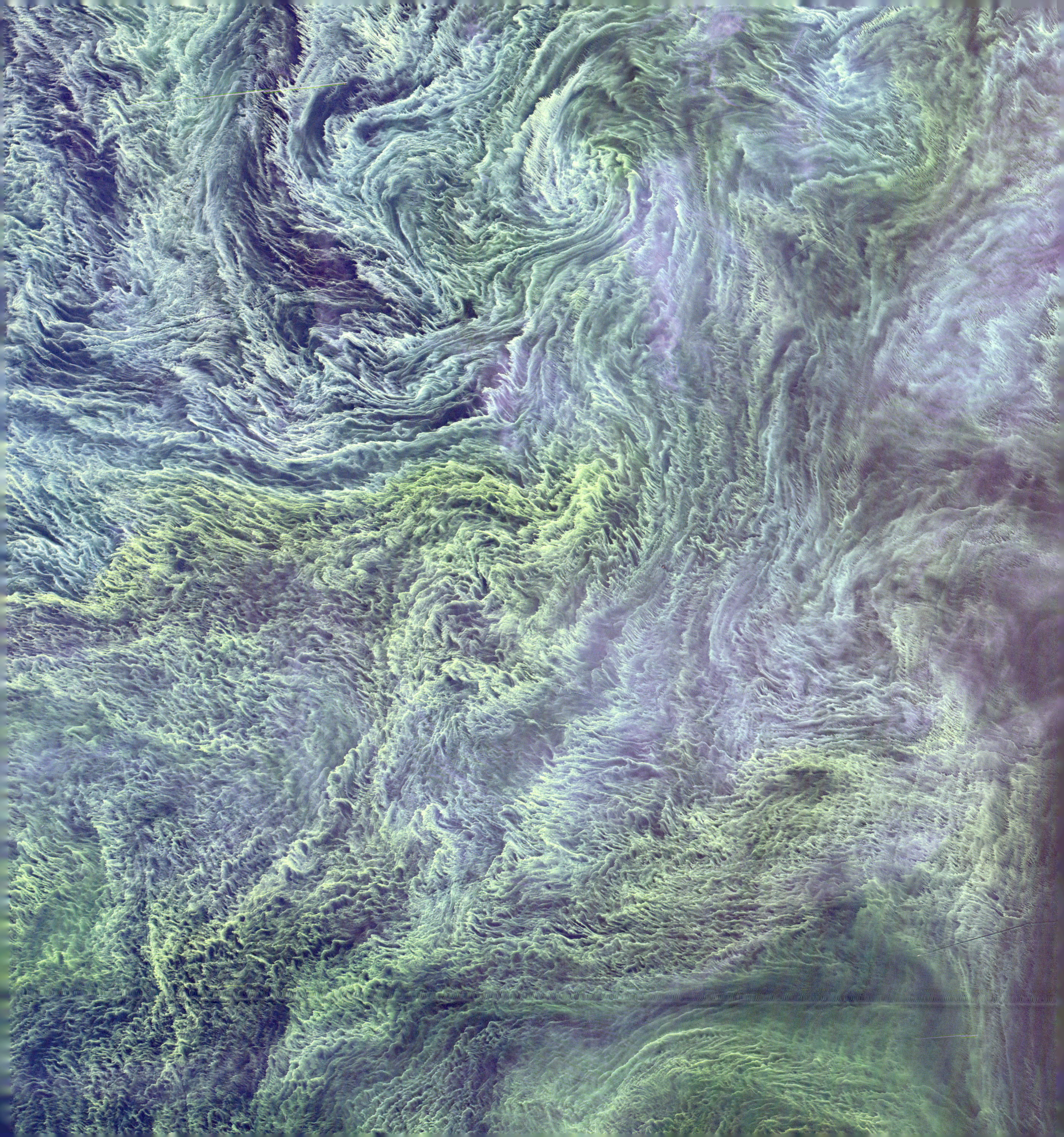

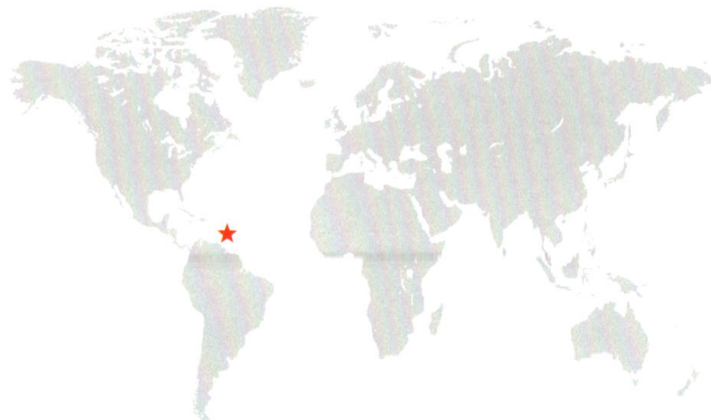

Waves Beneath the Waves Trinidad

Internal waves are the surface manifestation of slow waves that move tens of meters beneath the sea surface. These waves beneath the waves produce enough of an effect on the sea surface to be visible from space when they are enhanced by the reflection of sunlight, or sunglint, back toward a camera.

This January 2013 photograph from the International Space Station shows at least three sets of internal waves interacting. The most prominent set (top left) shows several waves moving from the northwest due to the tidal flow toward the north coast of Trinidad. Two less prominent sets can be seen further out to sea. All of these internal waves are probably caused by the shelf break near Tobago. The shelf break is the step between shallow seas (around continents and islands) and the deep ocean. It is the line at which tides usually start to generate internal waves.

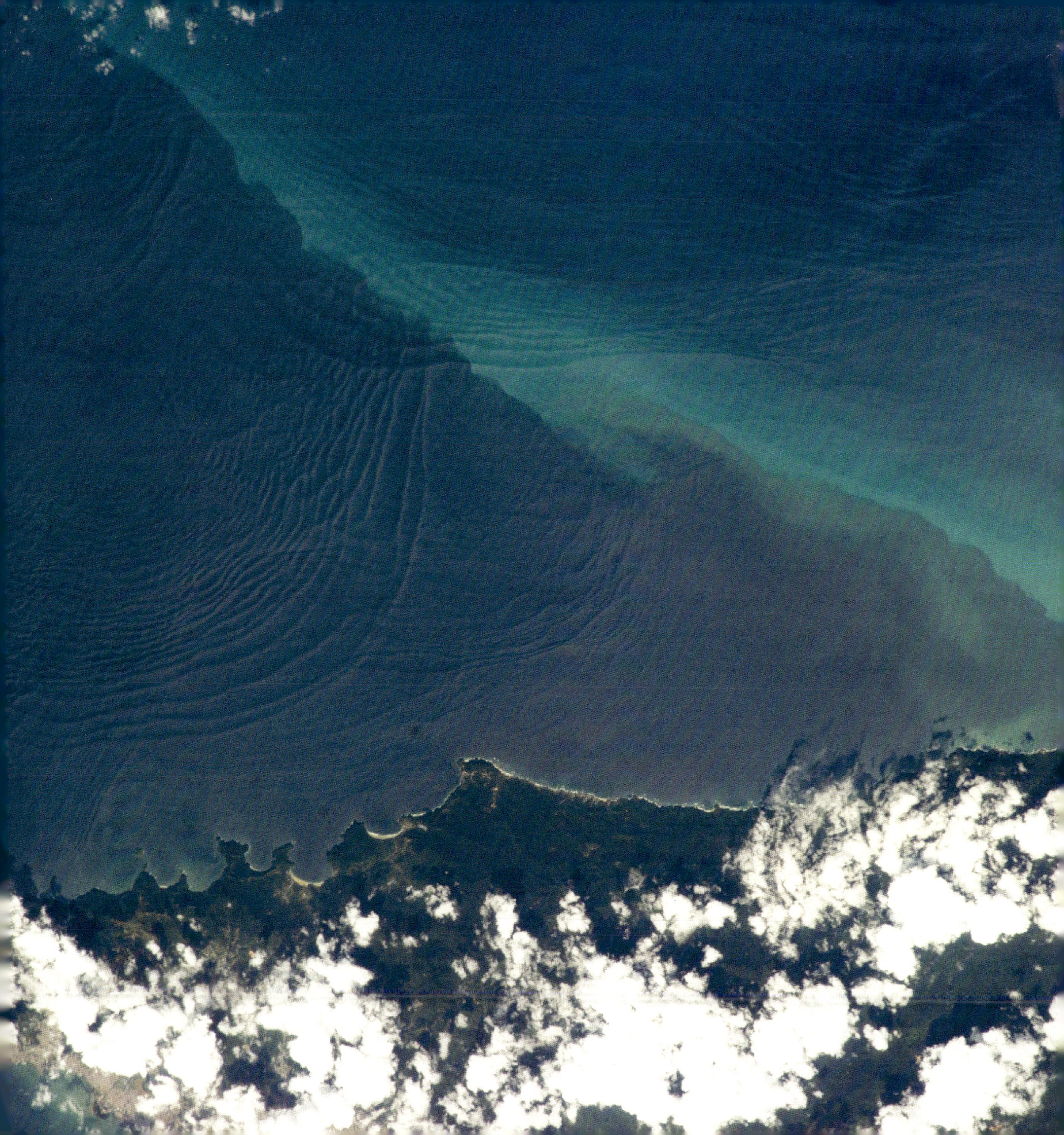

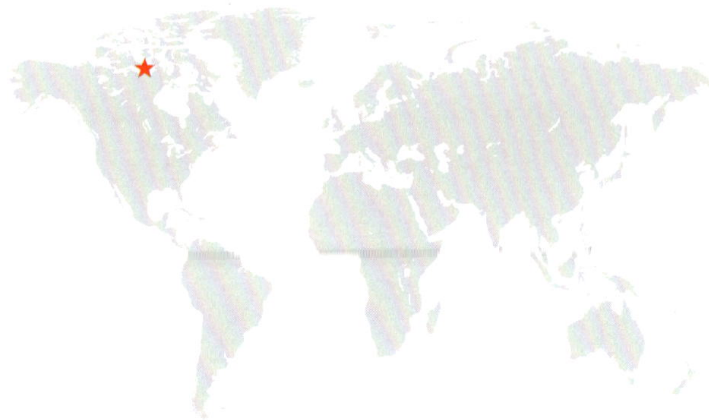

Land of Lakes
Canada

During the last Ice Age, nearly all of Canada was covered by a massive ice sheet. Thousands of years later, the landscape still shows the scars of that icy earth-mover. Surfaces that were scoured by retreating ice and flooded by Arctic seas are now dotted with millions of lakes, ponds, and streams. In this false-color view from the Terra satellite, water is various shades of blue, green, tan, and black, depending on the amount of suspended sediment and phytoplankton; vegetation is red.

The region of Nunavut Territory is sometimes referred to as the "Barren Grounds," as it is nearly treeless and largely unsuitable for agriculture. The ground is snow-covered for much of the year, and the soil typically remains frozen (permafrost) even during the summer thaw. Nonetheless, this July 2001 image shows plenty of surface vegetation in midsummer, including lichens, mosses, shrubs, and grasses. The abundant fresh water also means the area is teeming with flies and mosquitoes.

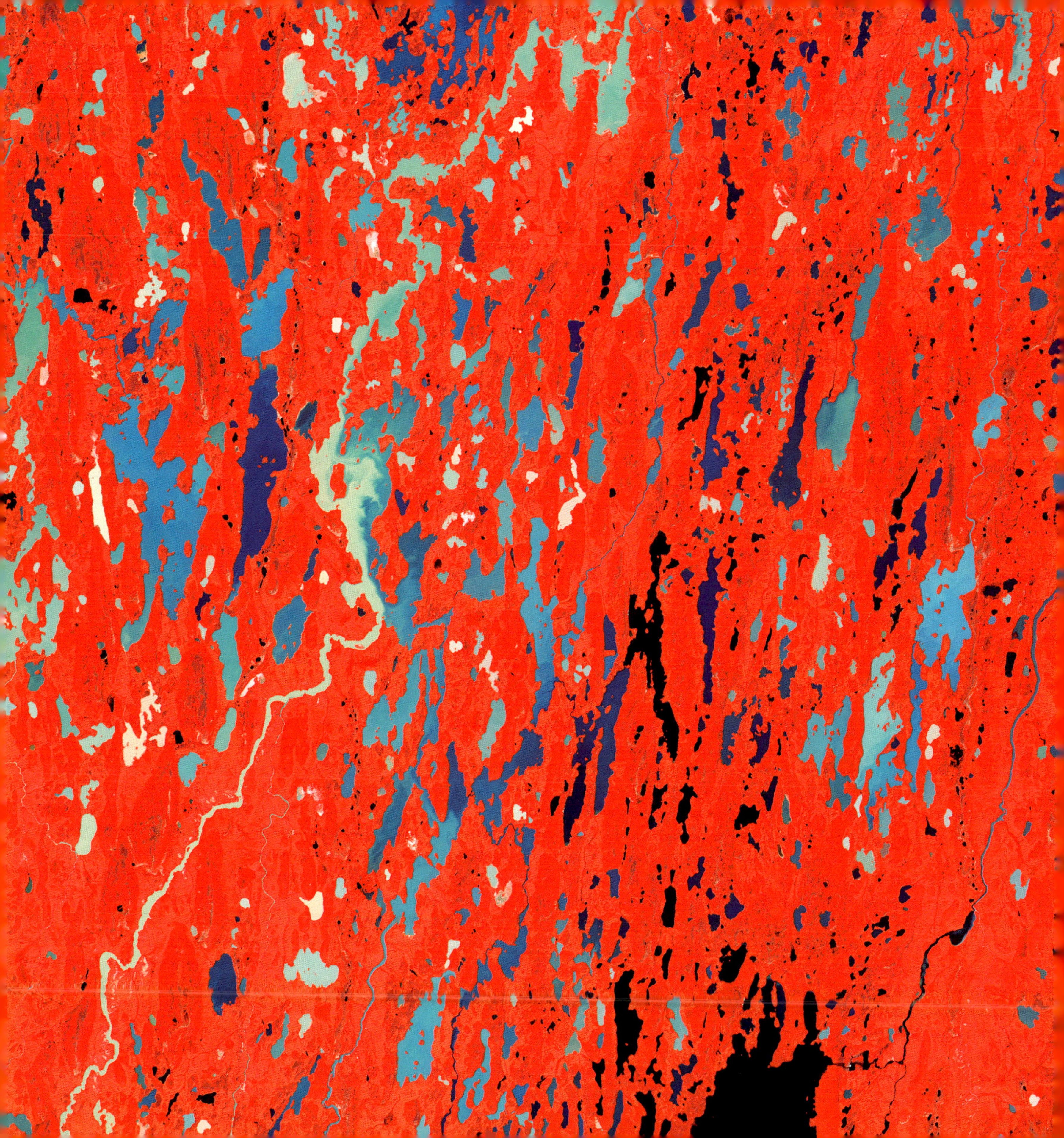

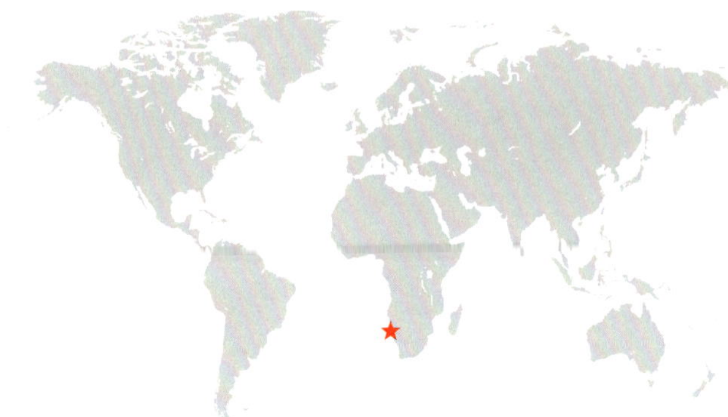

Plankton and Sulfur
Namibia

Off the coast of Namibia, the Benguela Current flows north and west from South Africa. It is enriched by iron and other nutrients from the Southern Ocean and from dust blowing off African coastal deserts. Easterly winds push surface waters offshore and promote upwelling near the coast, which brings up cold, nutrient-rich waters from the deeper ocean. These interactions can make the ocean come alive with color.

Bacteria in oxygen-depleted bottom waters consume organic matter and produce large amounts of hydrogen sulfide. As that gas bubbles up into more oxygen-rich water, the sulfur precipitates out and floats near the surface in yellow-green patches.

Further offshore, milky green water may be a bloom of phytoplankton. As these organisms consume sunlight and nutrients, they also consume oxygen and sometimes deplete it from the water. At the same time, those oxygen-depleted waters help sulfur-producing bacteria thrive.

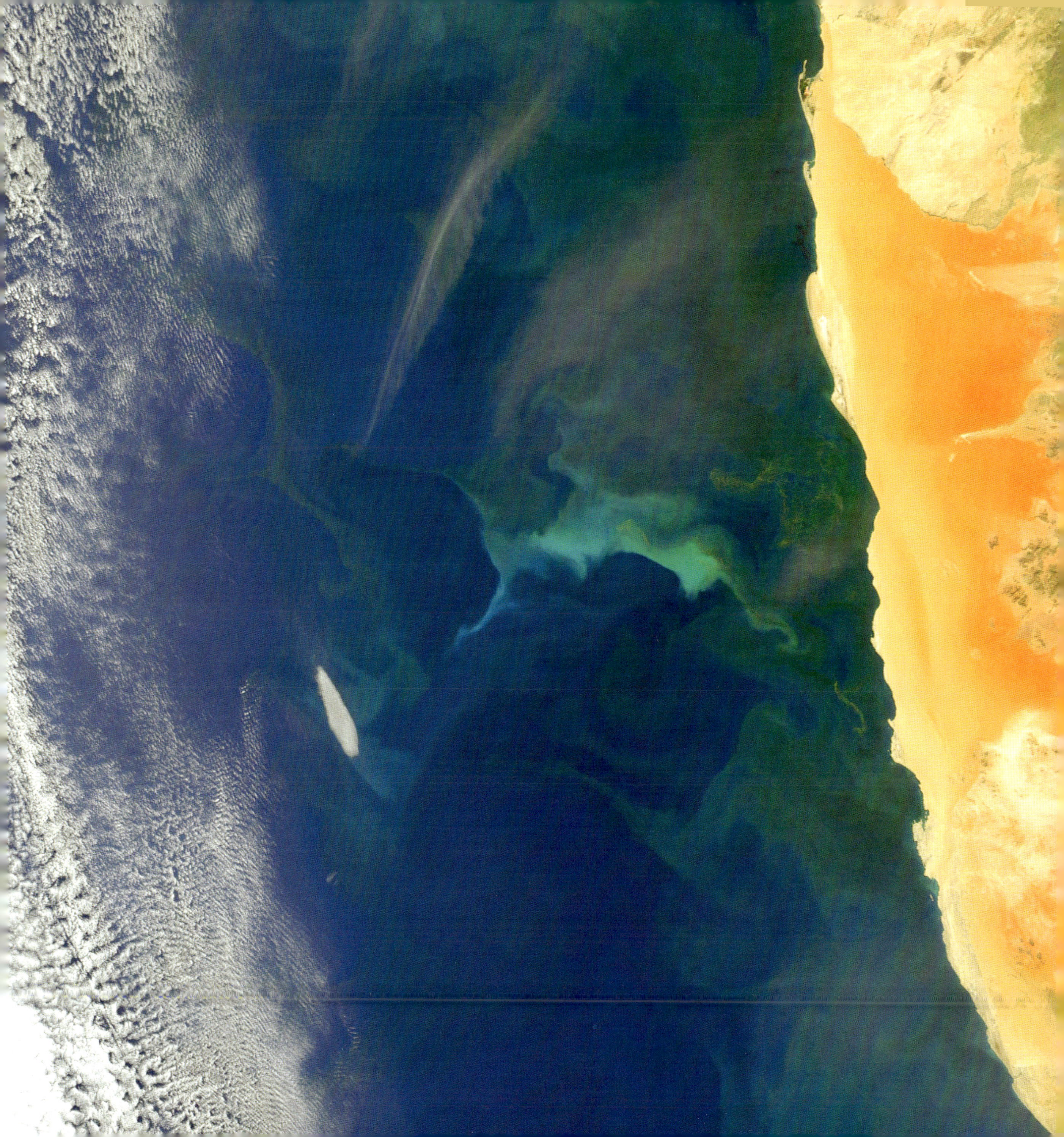

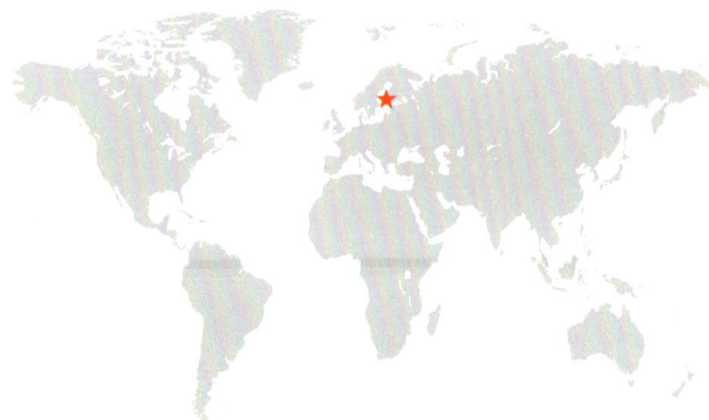

Åland Islands
Scandinavia

The Åland Islands lie in the Gulf of Bothnia, between Sweden and Finland. The archipelago consists of several large islands and roughly 6,500 small isles, many of them too small for human habitation. They are covered by pine and deciduous forest, meadows, and farmed fields. The region's characteristic red rapakivi granite also stands out.

These rocks were formed during the Proterozoic Eon, hundreds of millions of years before the dinosaurs. Massive ice sheets later sculpted the landscape, but these days people cut and use the granite in buildings and pavement. Landsat 5 acquired this image in June 2011.

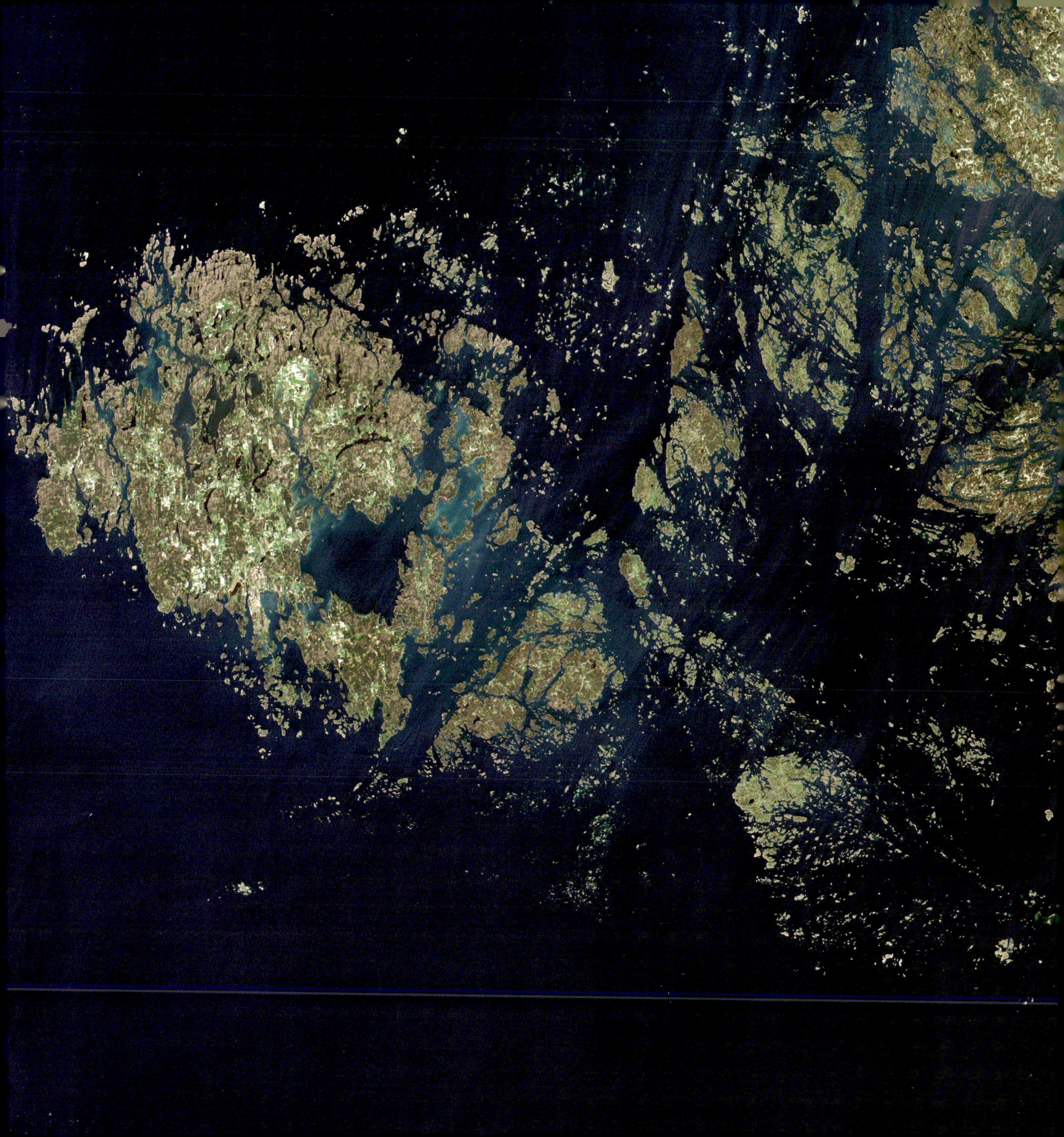

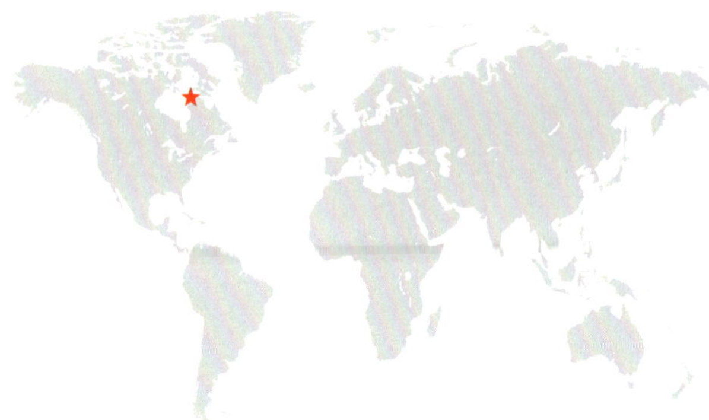

Crater Lakes with Clear Water
Canada

About 290 million years ago, two large asteroids smashed into Earth. The massive craters they left behind—now the Clearwater Lakes, or Lac à l'Eau Claire—are still visible from space.

When they struck, the binary asteroids crashed into a part of Earth's crust that was fairly close to the Equator. Since then, millions of years of plate tectonics have pushed the craters northward into what is now northwestern Quebec. A mere 20,000 years ago, massive ice sheets advanced and retreated, scouring the land of soil and rock during cool periods and then carving deep channels and rinsing the landscape with meltwater during warmer periods. The erosion was so complete that many land surfaces were scraped down to underlying bedrock, exposing some of the oldest rocks in the world. Meanwhile, the meltwater from retreating glaciers left a dense network of linear lakes and streams that now dominate the surface.

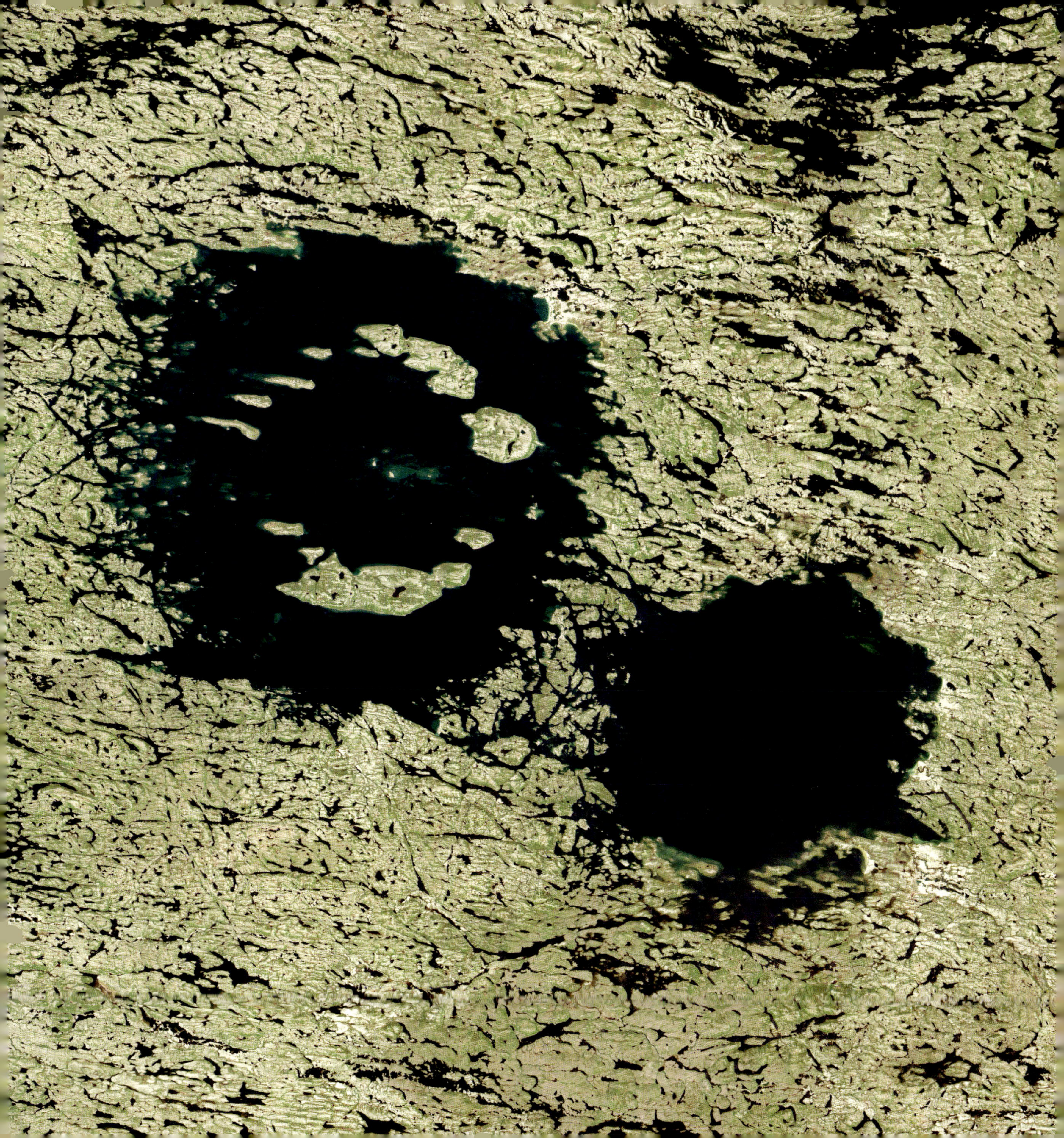

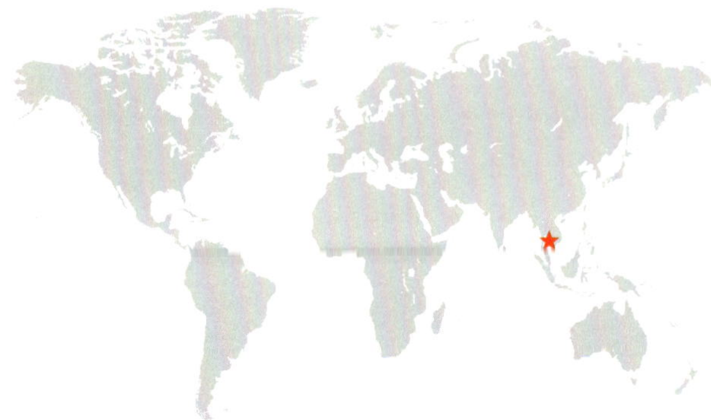

Mergui Archipelago
Southeast Asia

Near the border of Burma (Myanmar) and Thailand, more than 800 islands rise amid extensive coral reefs in the Andaman Sea. This is the Mergui Archipelago.

Captain Thomas Forrest of the East India Company first reported on the region to Europeans after a 1782 expedition, describing islands inhabited by a nomadic fishing culture. These people, known as the Moken, still call the archipelago home and mostly live a hunter-gatherer lifestyle. The small population of the archipelago has helped preserve its high diversity of plants and animals, making it a compelling travel spot for ecotourism—both above and below the water line.

In this view of Auckland Bay and Whale Bay, white swirling patterns in the near-shore waters are sediments that are carried out by rivers and deposited on the seafloor.

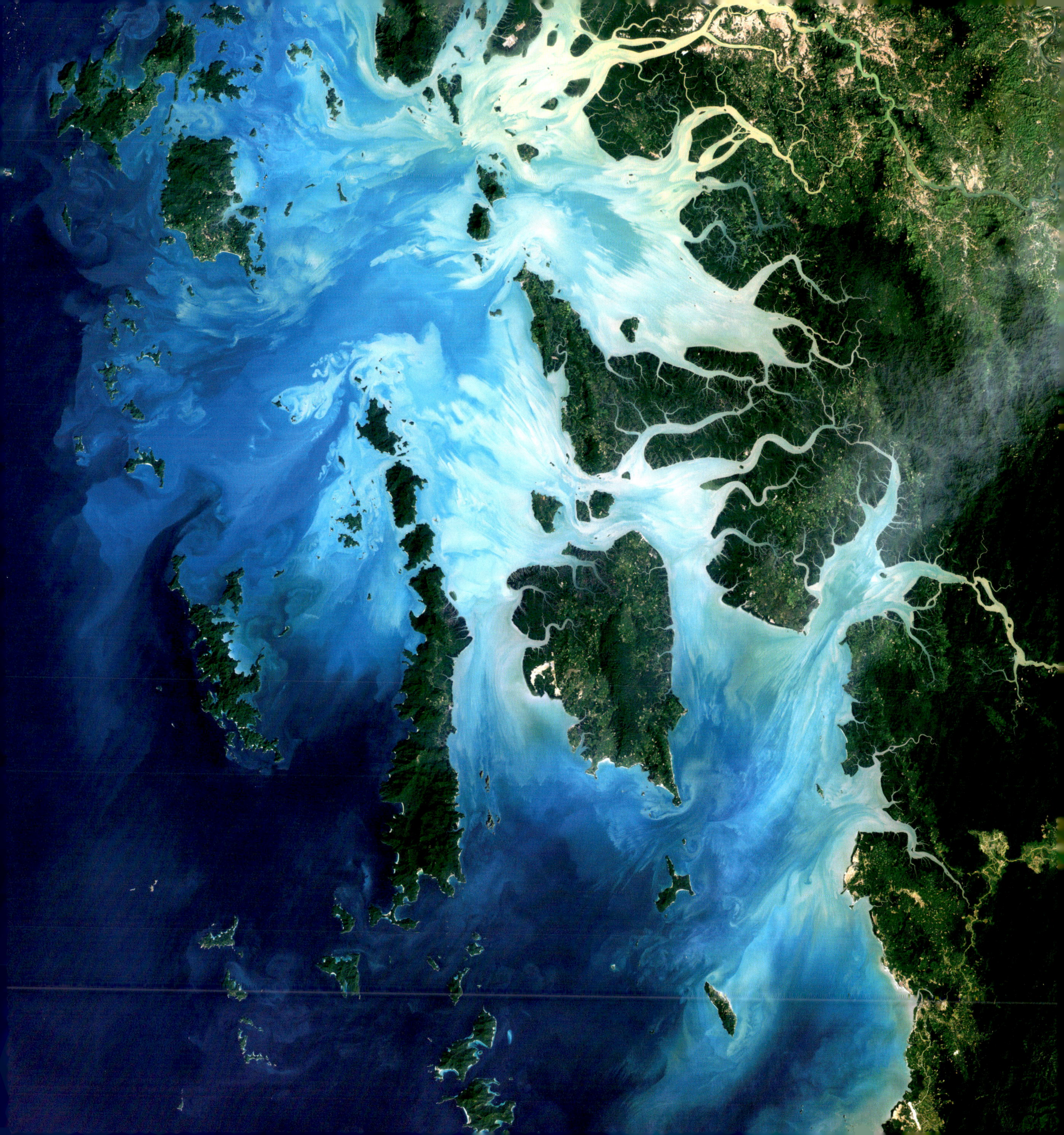

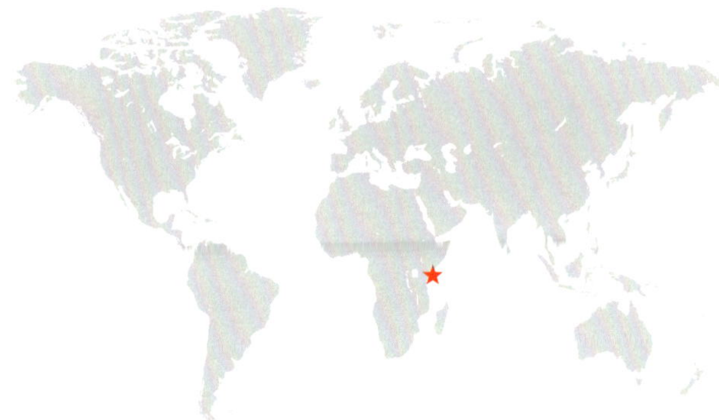

Scarlet Lake Natron
Tanzania

Lake Natron is mostly inhospitable to life, but it is gorgeous to the eye. The lake in Tanzania receives less than 500 millimeters (20 inches) of rain in most years. Evaporation usually exceeds that amount, and the lake needs input from some local rivers to maintain a water supply in the dry season.

This Landsat 8 image from March 2017 shows Lake Natron's chromatic charisma. Volcanism helps make the unusual color. Nearby volcanoes produce molten mixtures of sodium carbonate and calcium carbonate salts that move through faults and well up in hot springs. This briny, alkaline environment is too harsh for most common types of life, but salt-loving microorganisms (haloarchaea) bloom in the shallow pools and impart pink and red colors to the water.

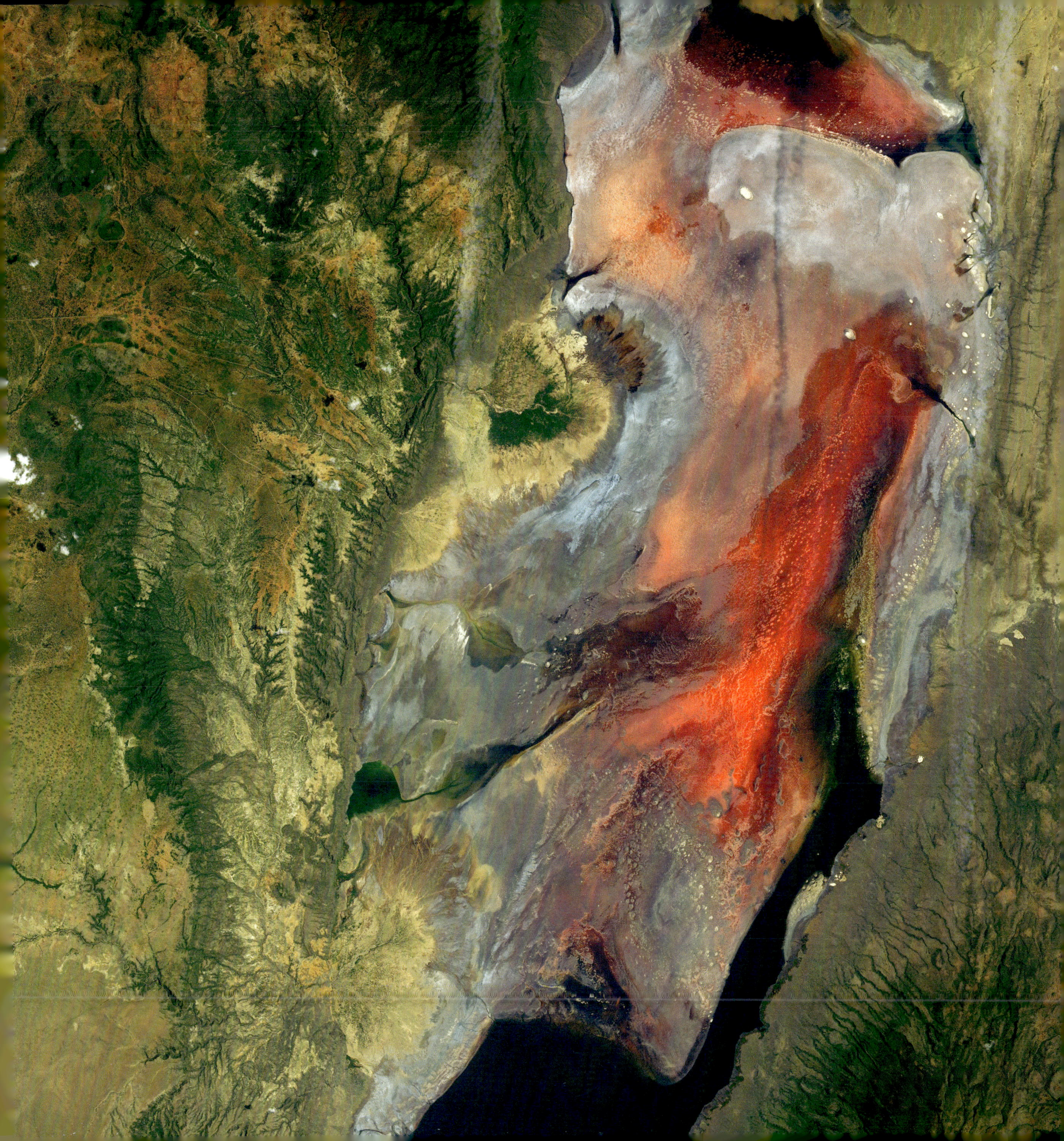

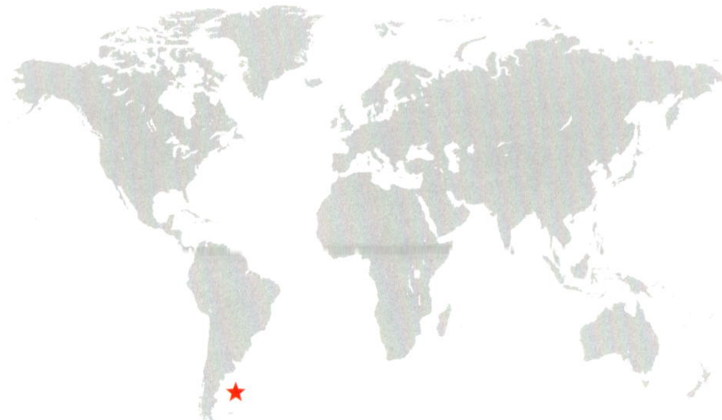

Swirling Bloom off Patagonia Argentina

Interesting art often springs out of the convergence of different ideas and influences. And so it is with nature.

Off the coast of Argentina, two strong ocean currents converge and often stir up a colorful brew, as shown in this Aqua image from December 2010.

This milky green and blue bloom formed on the continental shelf off of Patagonia, where warmer, saltier waters from the subtropics meet colder, fresher waters flowing from the south. Where these currents collide, turbulent eddies and swirls form, pulling nutrients up from the deep ocean. The nearby Rio de la Plata also deposits nitrogen- and iron-laden sediment into the sea. Add in some midsummer sunlight, and you have a bountiful feast for microscopic, floating plants known as phytoplankton, which form the center of the ocean food web.

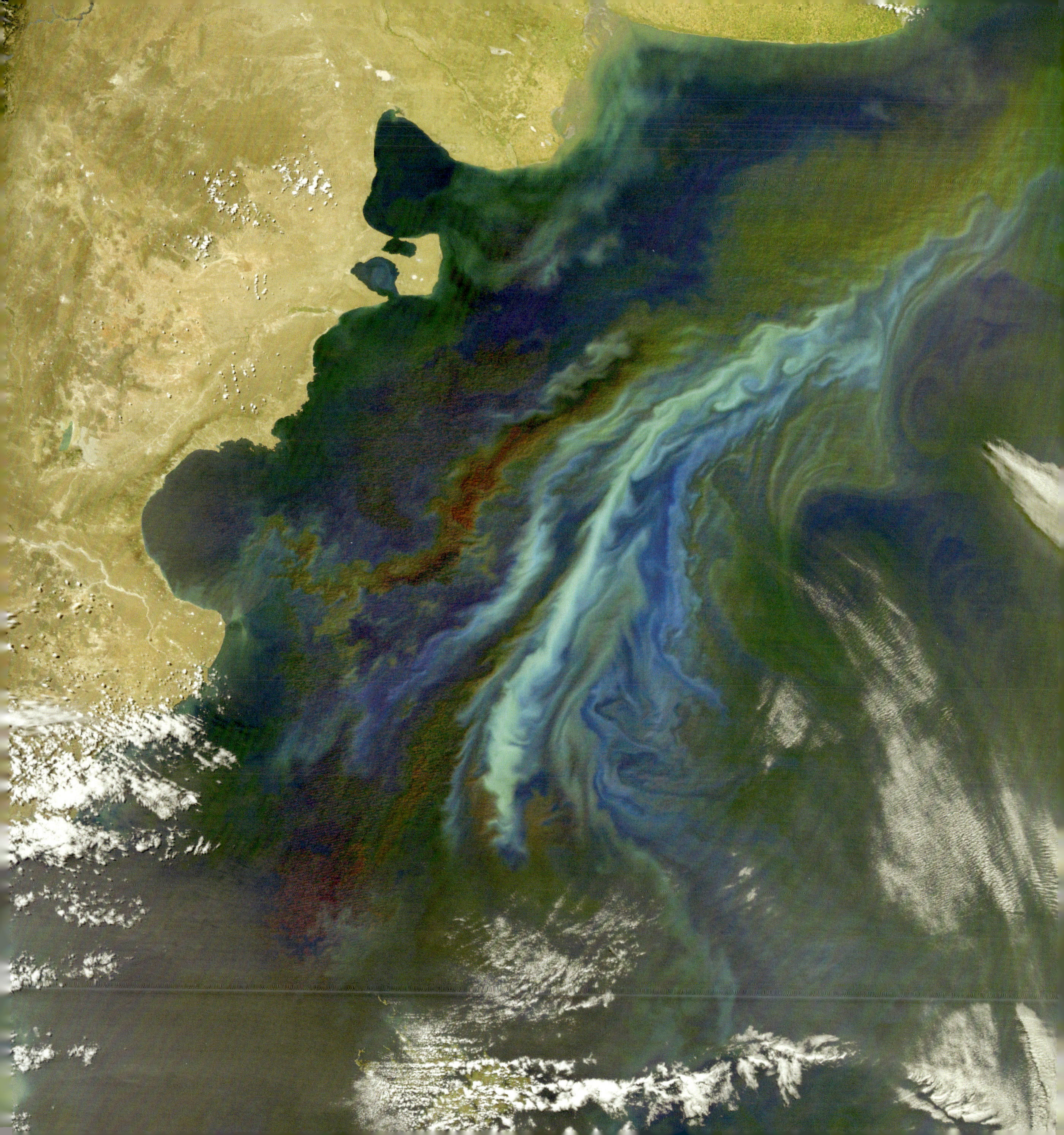

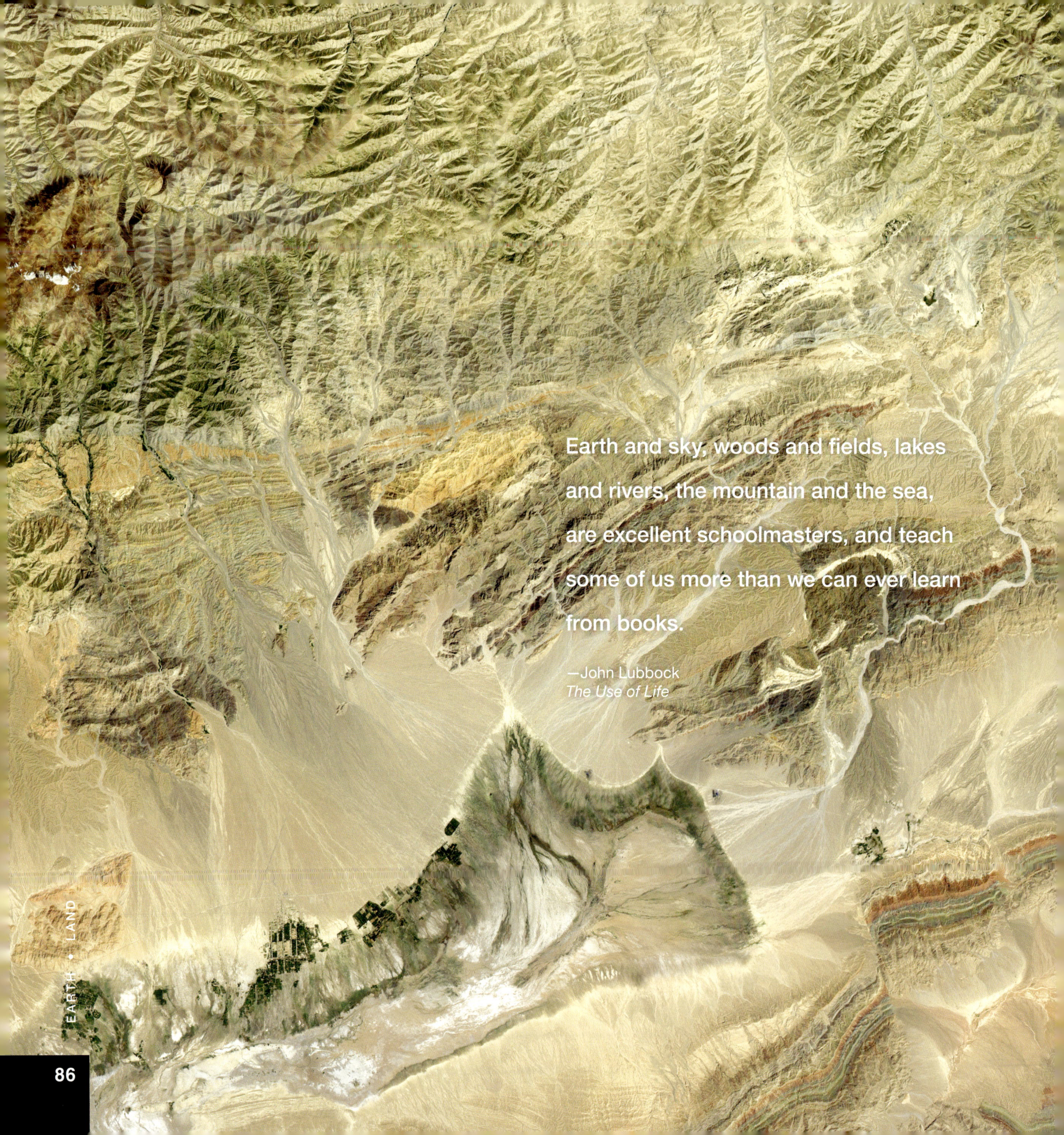

Earth and sky, woods and fields, lakes and rivers, the mountain and the sea, are excellent schoolmasters, and teach some of us more than we can ever learn from books.

—John Lubbock
The Use of Life

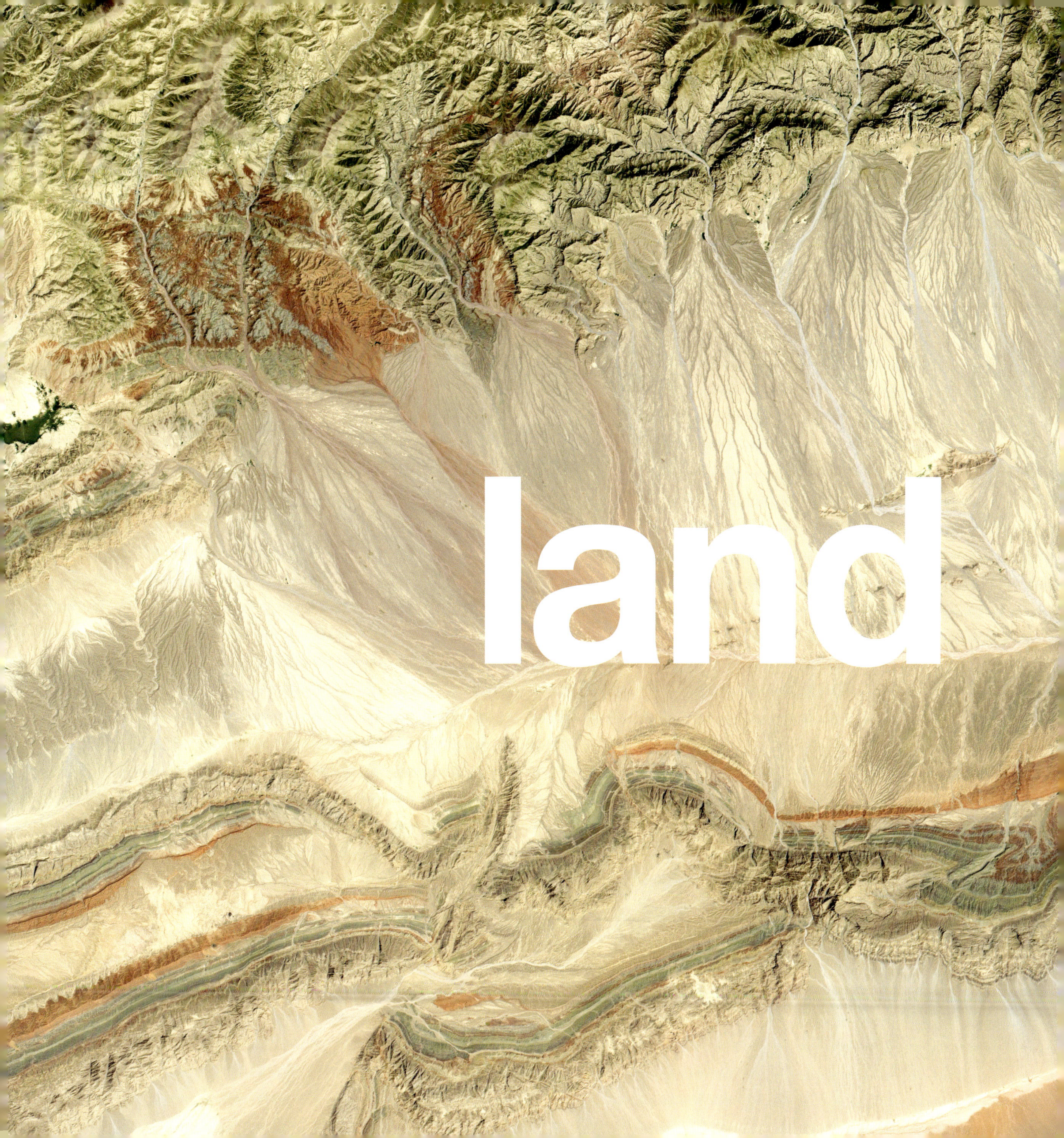
land

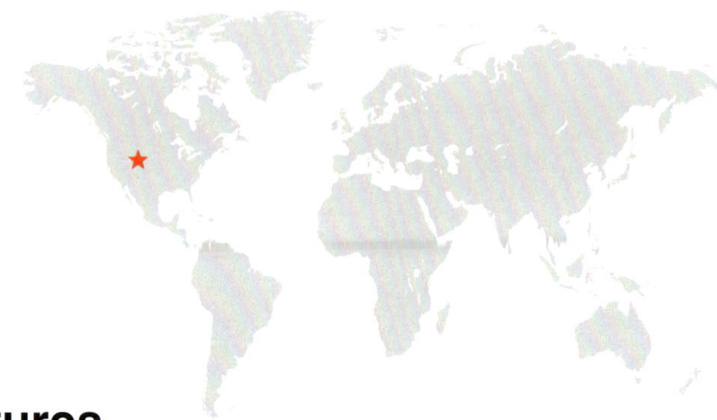

A Curious Ensemble of Wonderful Features
United States

When John Wesley Powell led an expedition down the Colorado River and through the Grand Canyon in 1869, he was confronted with a daunting landscape. At its highest point, the serpentine gorge plunged 1,829 meters (6,000 feet) from rim to river bottom, making it one of the deepest canyons in the United States. In just 6 million years, water had carved through rock layers that collectively represented more than 2 billion years of geological history.

"The wonders of the Grand Canyon cannot be adequately represented in symbols or speech," Powell wrote in his log. Powell was seeing the canyon mainly from river level; there was no technology that provided views of the landscape from space then. If there had been, he would have seen something similar to what Landsat 8 observed in March 2013.

The Colorado River traces a line across the arid Colorado Plateau. Treeless areas are beige and orange; green areas are forested. The river water is brown and muddy, a common occurrence in spring when melting snows cause water levels to swell and pick up extra sediment.

It took Powell months to navigate the gorge. By the time he had arrived in the area that is now Lake Mead (visible on the next page, far left), his men were weary and four had deserted. For water and sediment transported by the Colorado, the journey is much quicker; it takes just a handful of days.

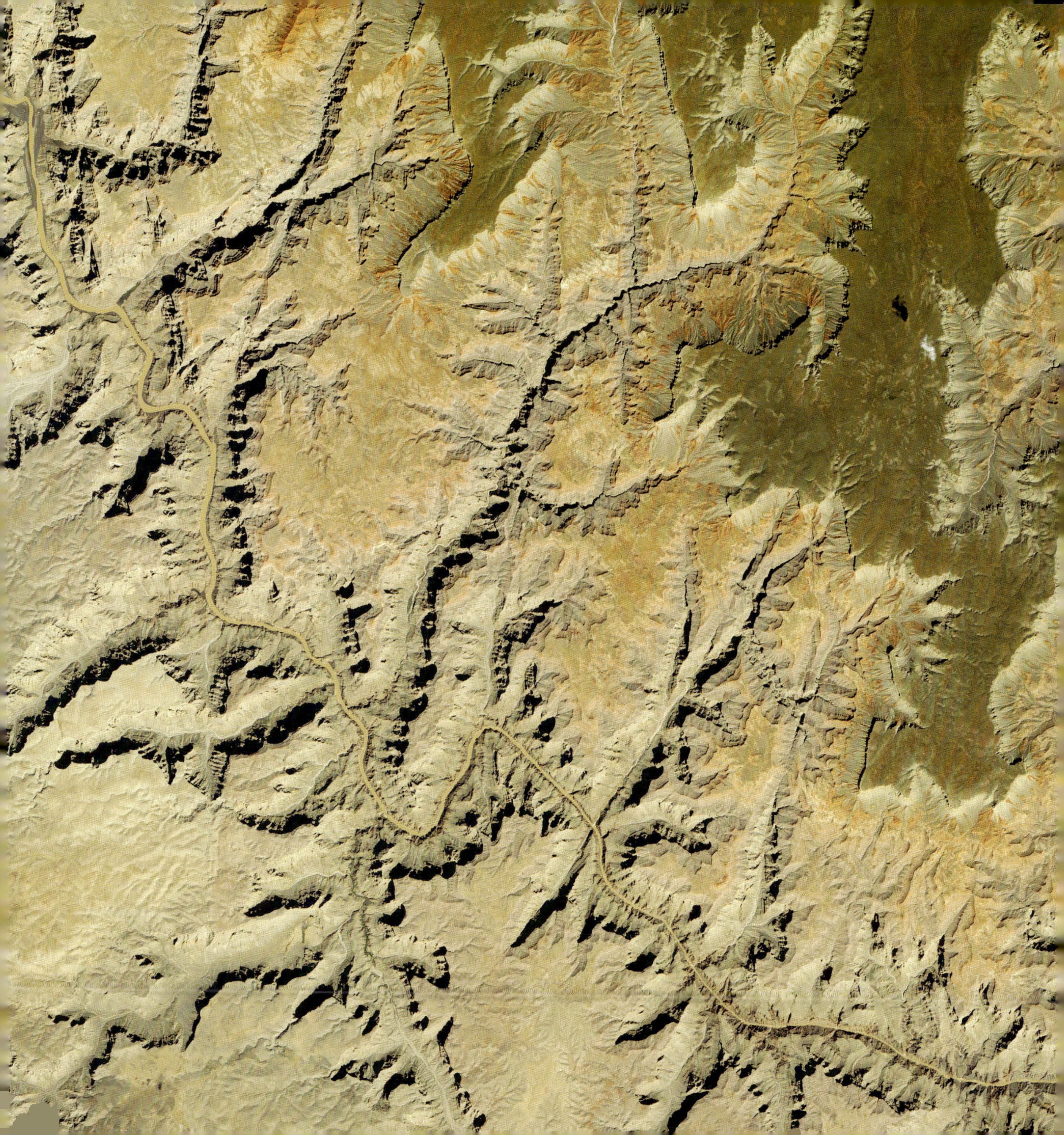

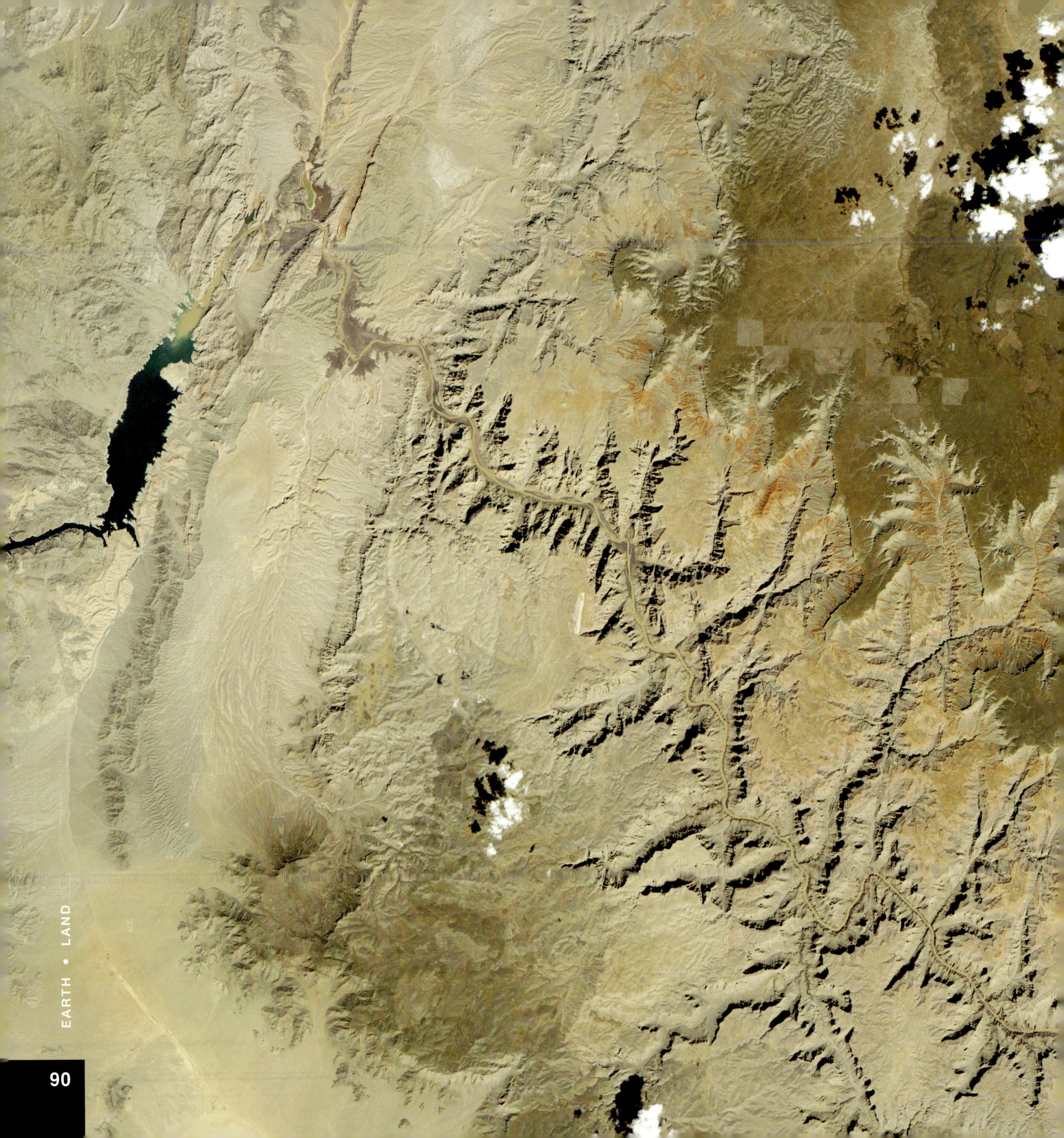

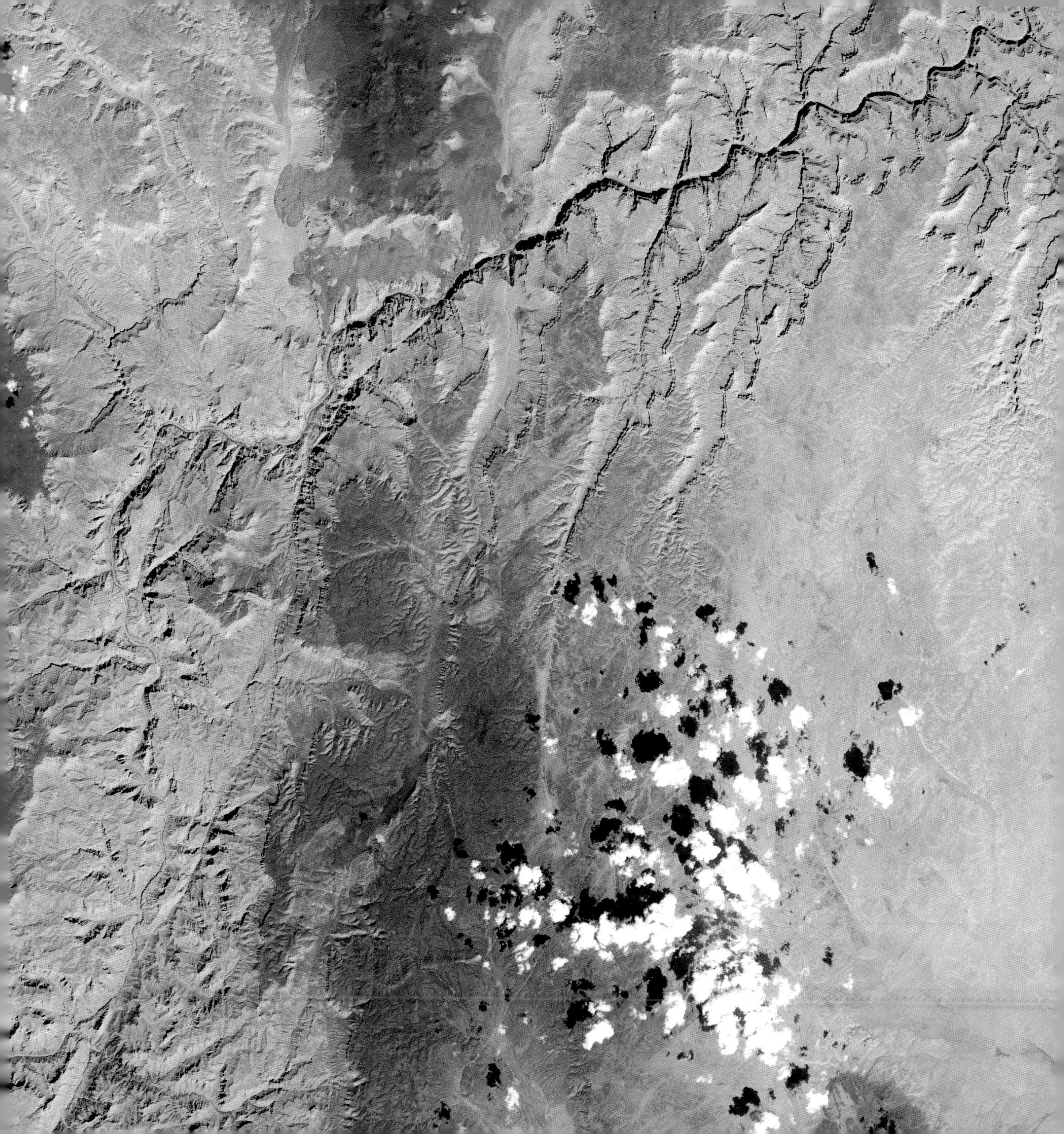

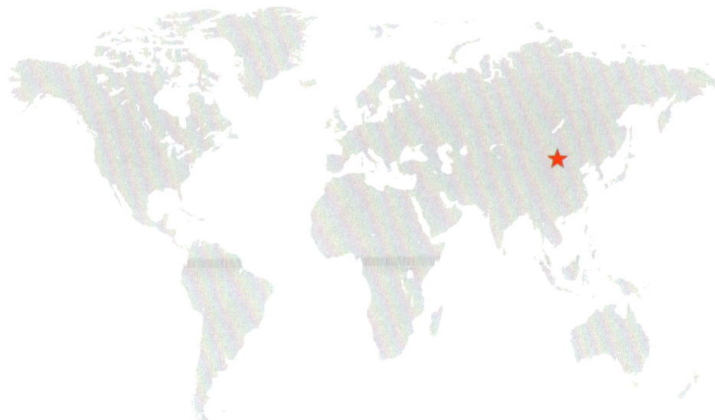

Megadunes and Desert Lakes
Mongolia

In the Badain Jaran, nearly 100 lakes mingle with the tallest sand dunes in the world. Researchers have long studied these features in China, yet mystery continues to enshroud them.

Situated in the Alxa Desert region of Inner Mongolia, the Badain Jaran naturally piles up megadunes towering 200 to 300 meters (650 to 1,000 feet) tall. Scientists are still puzzling over how the dunes grow so large, investigating a combination of wind patterns and underlying geology.

Another enigma is the source of water for the lakes. Scientists are working to figure out the relative contributions from precipitation, groundwater, snowmelt, and paleowater. Regardless of the water source, they know that some lakes have shrunk or disappeared in recent years.

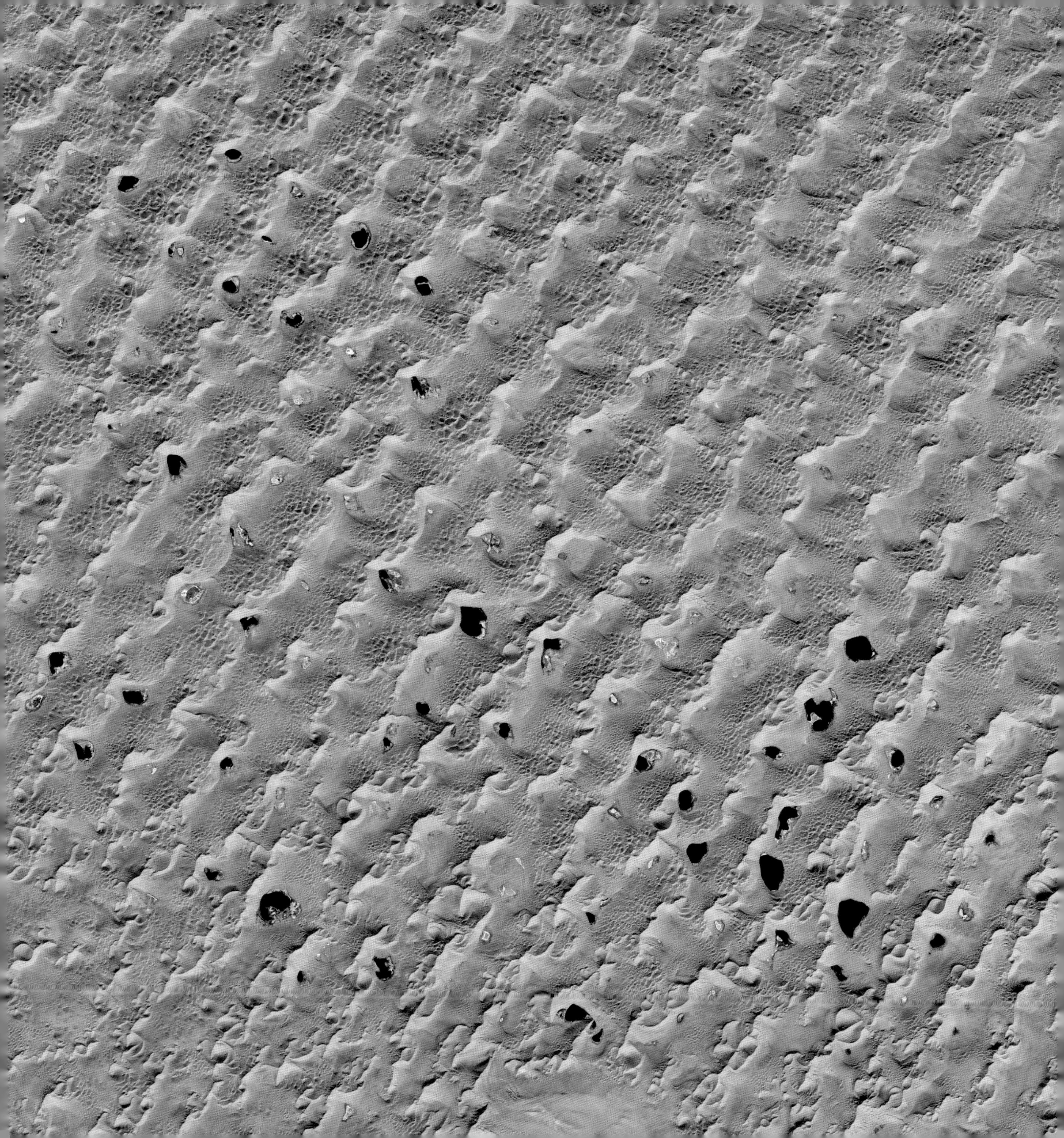

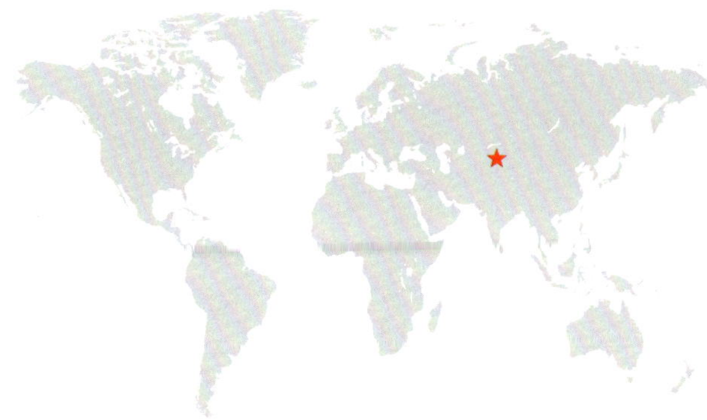

Colorful Faults of Xinjiang China

Just south of the Tien Shan mountains, in northwestern Xinjiang province, a remarkable series of ridges dominates the landscape. The hills are decorated with distinctive red, green, and cream-colored sedimentary rock layers. The colors reflect rocks that formed at different times and in different environments. The red layers near the top of the sequence are Devonian sandstones formed by ancient rivers. The green layers are Silurian sandstones formed in a moderately deep ocean. The cream-colored layers are Cambrian-Ordovician limestone formed in a shallow ocean.

Landsat 8 captured this image of the Keping Shan thrust belt in July 2013. When land masses collide, the pressure can create what geologists call "fold and thrust belts." Slabs of sedimentary rock that were laid down horizontally can be squeezed into wavy anticlines and synclines. Sometimes the rock layers break completely, and older layers of rock pile up on top of younger layers.

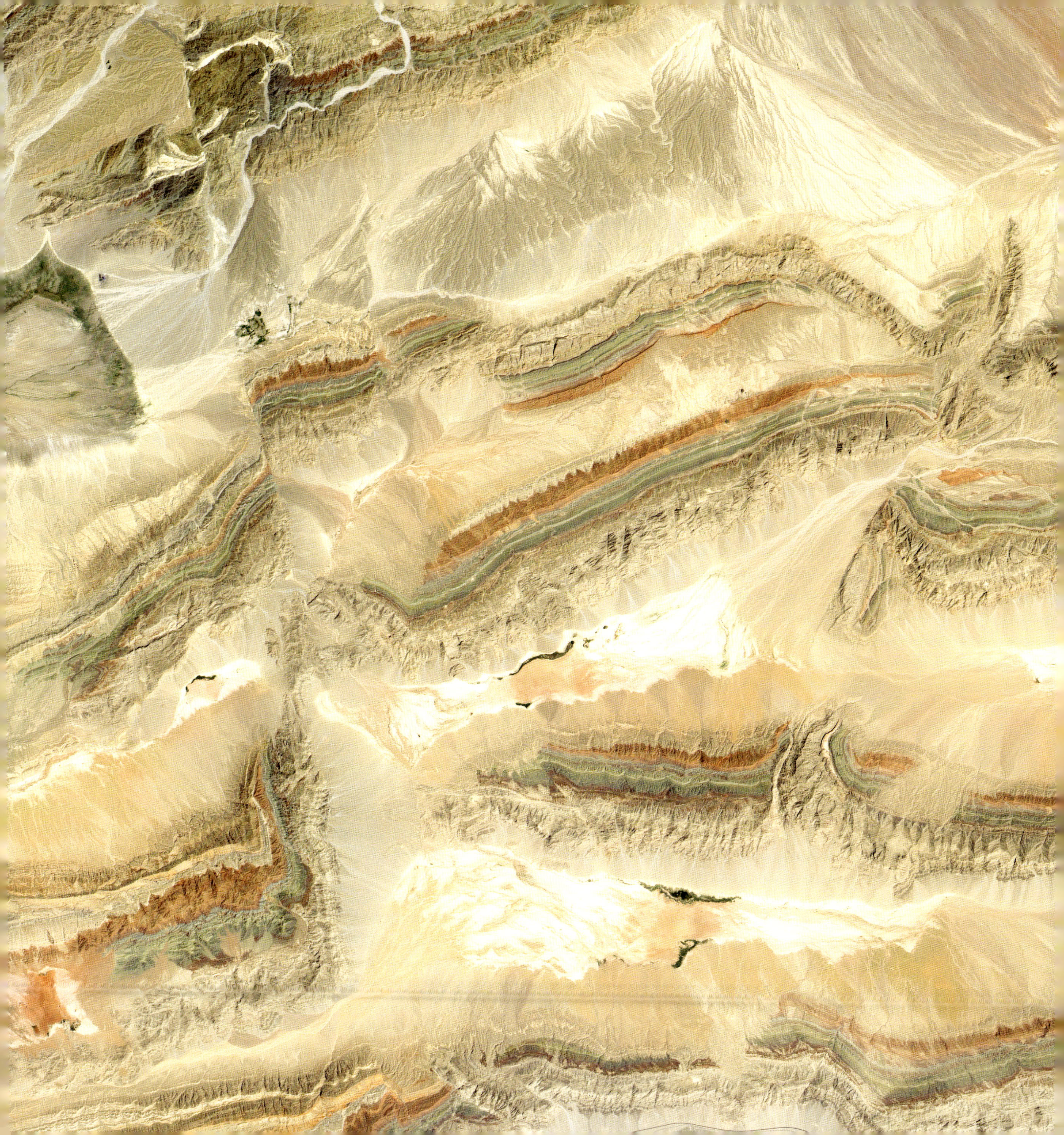

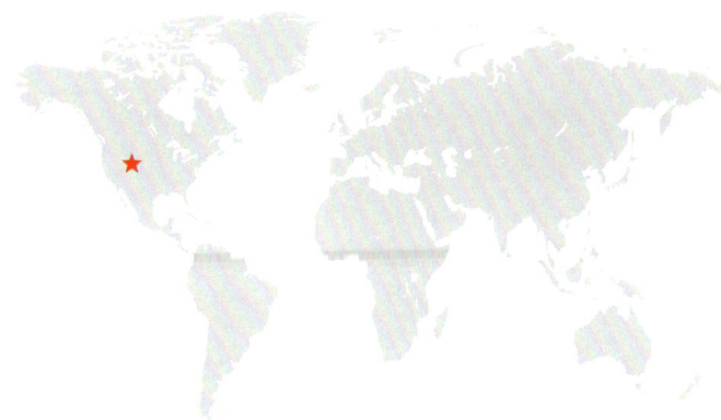

Bowknot Bend
United States

This section of the Green River canyon in eastern Utah is known as Bowknot Bend because of the way the river doubles back on itself. The loop carries river rafters 14.5 kilometers (9 miles) before bringing them back to nearly the same point they started from—though on the other side of a low, narrow saddle. The reason for the tight bends in the Green River is the same as it is for the mighty Mississippi: river courses often wind over time when they flow across a bed of relatively soft sediment in a floodplain.

In this January 2014 photograph taken from the International Space Station, the Green River appears dark because it lies in deep shadow, 300 meters (1,000 feet) below the surrounding landscape. The yellow-tinged cliffs that face the rising Sun give a sense of the steep canyon walls. The straight white line across the scene is the contrail from a jetliner.

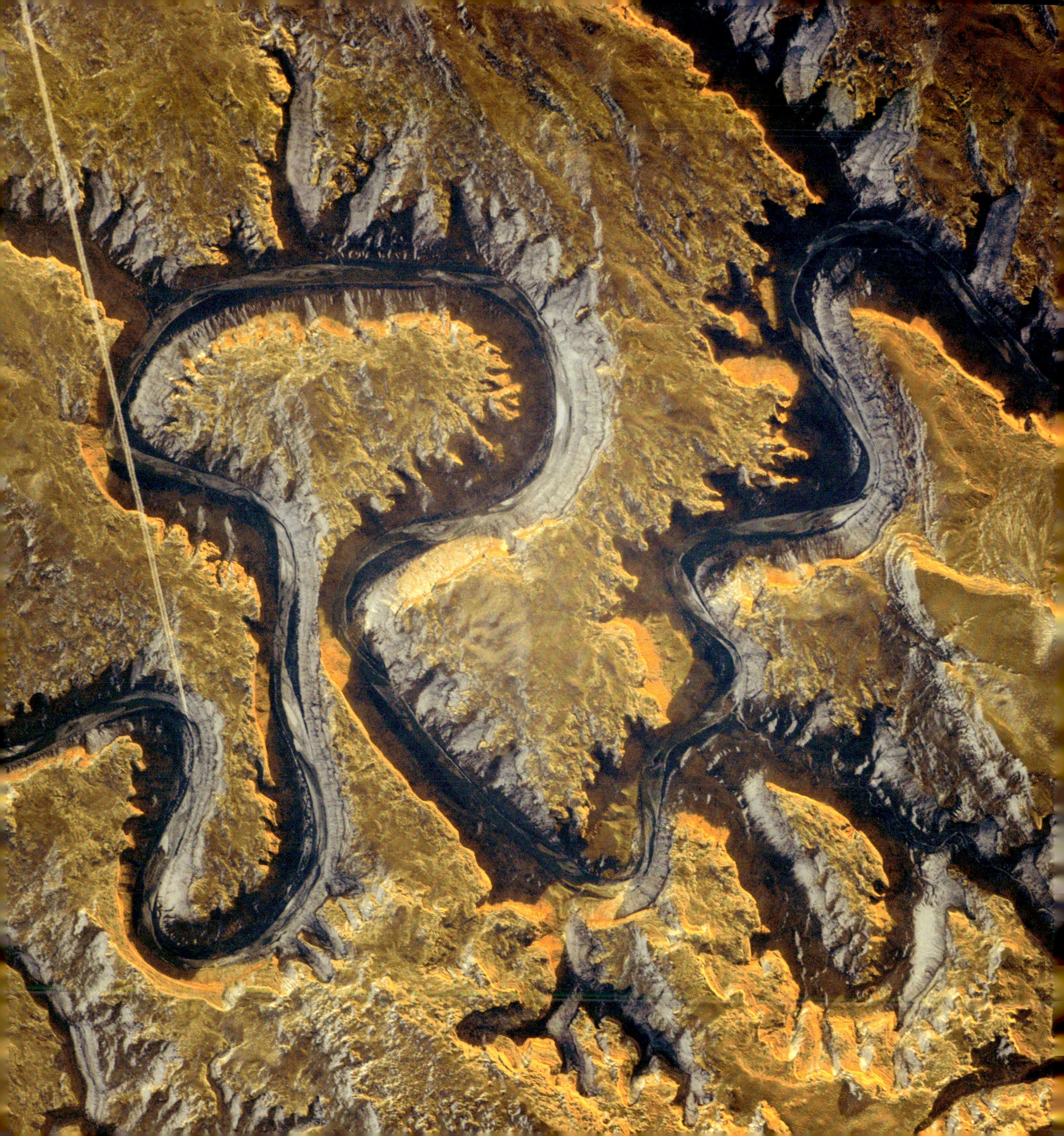

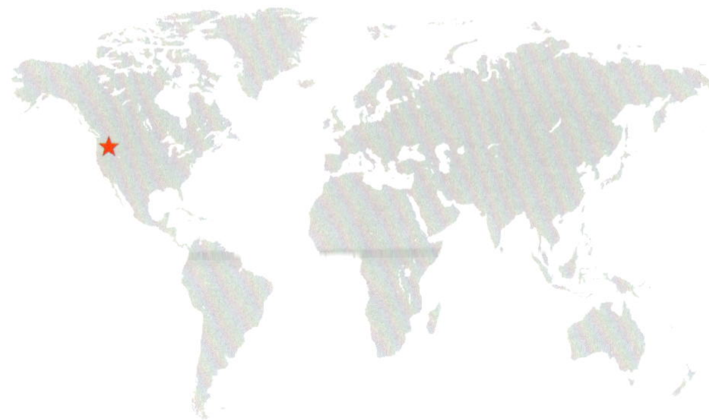

From Rainforest to Rain Shadow
United States

Within a three-hour drive across Oregon, you can visit a beach, a temperate rainforest, a mountain glacier, and the high desert. The diversity of the landscape is mostly driven by the interaction of air masses and mountains.

This false-color Landsat 5 image from October 2011 shows the bare soil and sparse vegetation of the high desert in shades of pink, together with the deep-green vegetation on the west side of the Cascade Mountains. The one blue spot is the glacial cap of Mount Hood.

The transition from green to brown is indicative of a "rain shadow." Winds blowing from the west carry moisture from the Pacific Ocean. As the air moves up into the mountains, it cools and the pressure decreases; the moisture condenses and falls out as rain or snow. On the eastern side, as the elevation drops, the air pressure increases and the air warms, effectively shutting off precipitation because the air can better hold the remaining moisture.

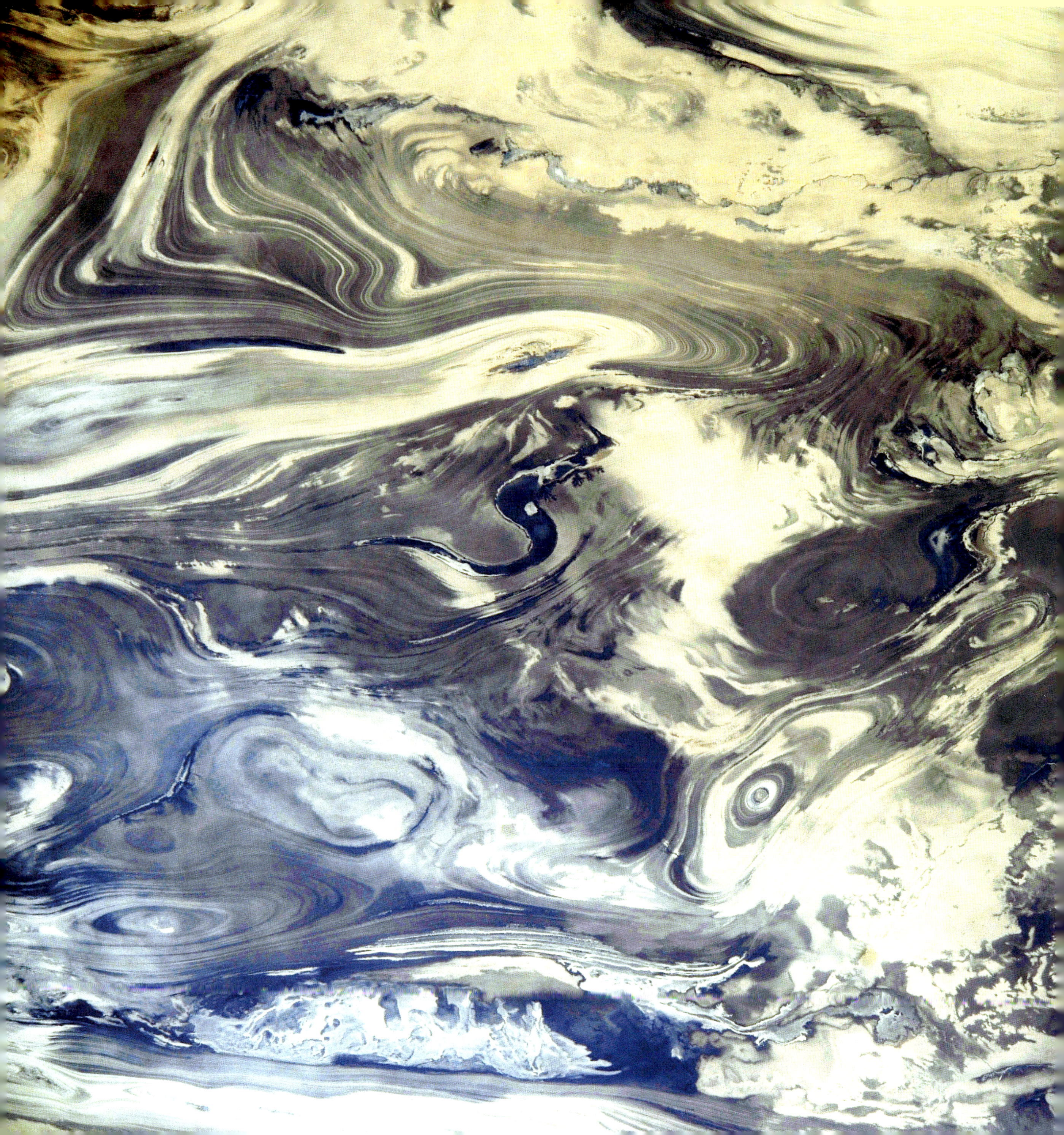

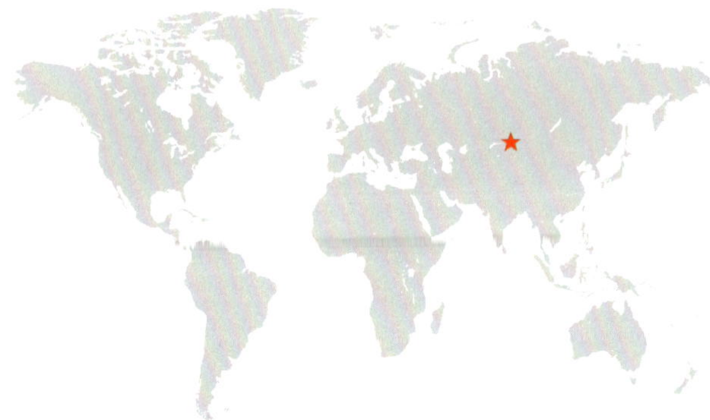

Fanning Out in Farmland
Kazakhstan

Mountain streams are usually confined to narrow channels and tend to transport large amounts of gravel, sand, clay, and silt—what geologists call alluvium. When such a stream pours onto a relatively flat valley or basin, it often spreads out to into multiple, interlacing channels. Over time, the channels migrate back and forth, creating fan-shaped deposits known as alluvial fans.

Landsat 8 captured this view of Kazakhstan's Almaty Province in September 2013. On the lower left, the Tente River flows through the foothills of the Dzungarian Alatau range. Where the Tente emerges, it spreads out and becomes a braided stream. The movement of the channel over time has left a large alluvial fan. In arid areas, these fans are often used for agriculture because they are relatively flat and provide groundwater for irrigation. The blocky green patterns show fields or pasture land.

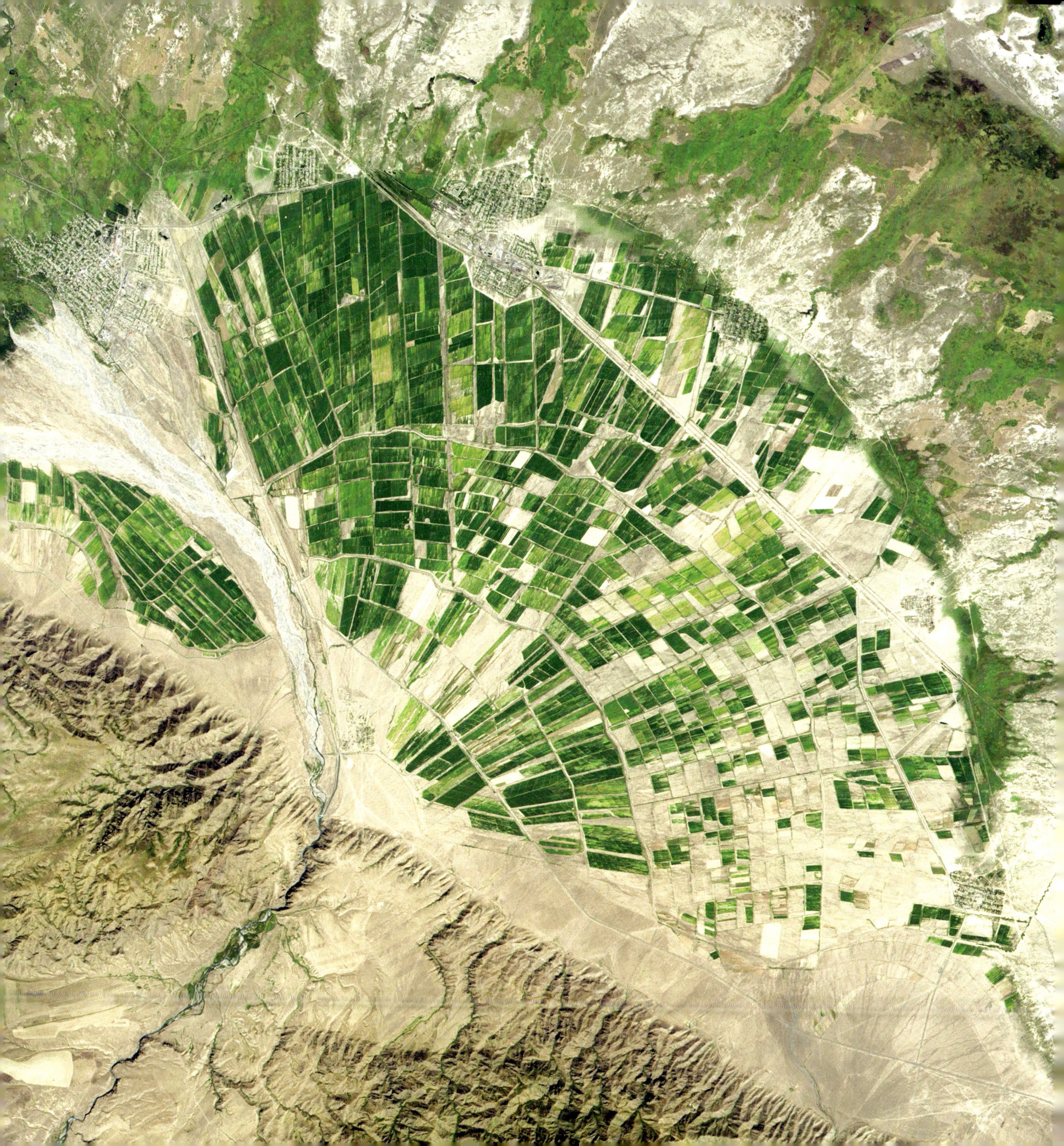

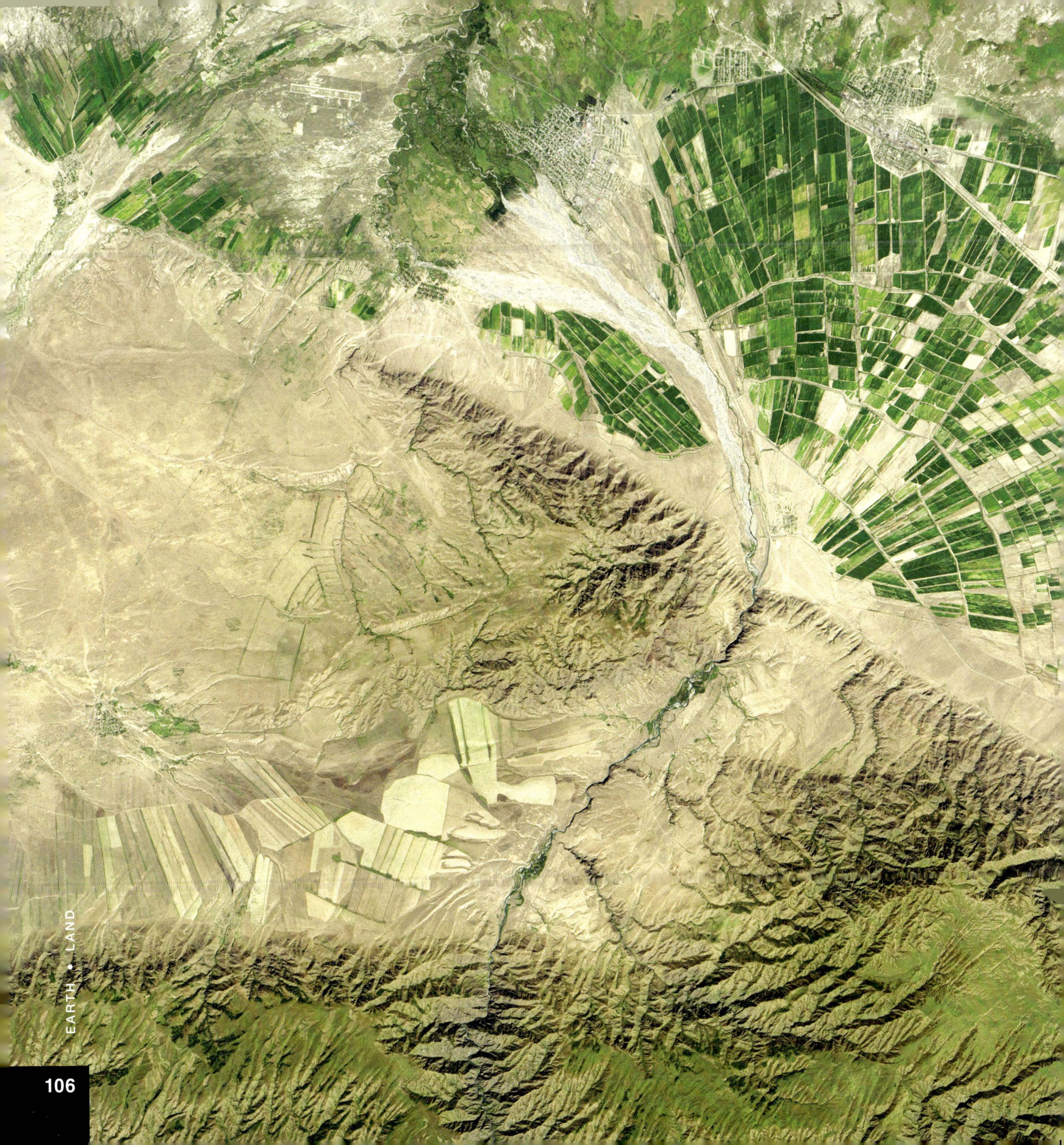

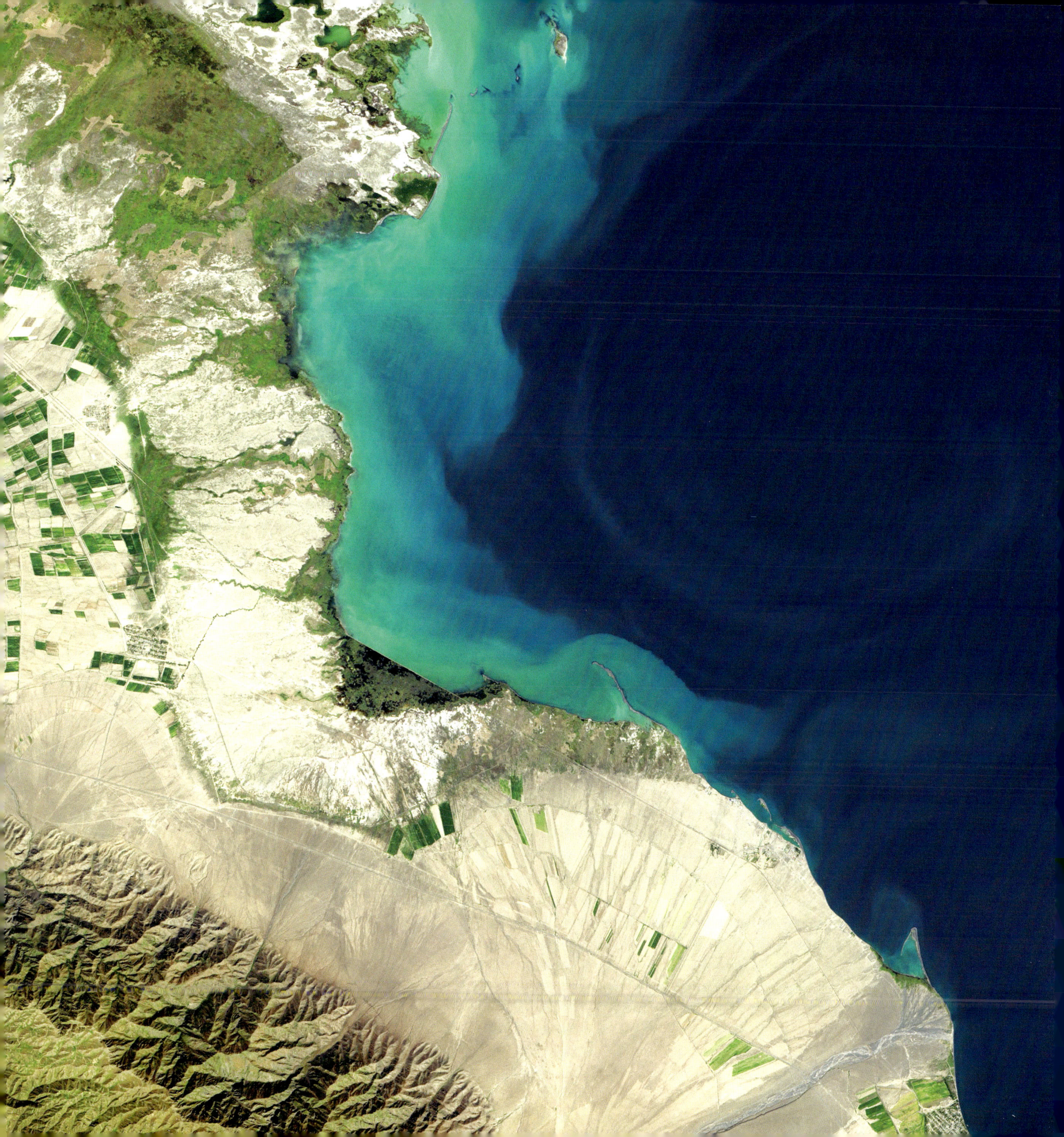

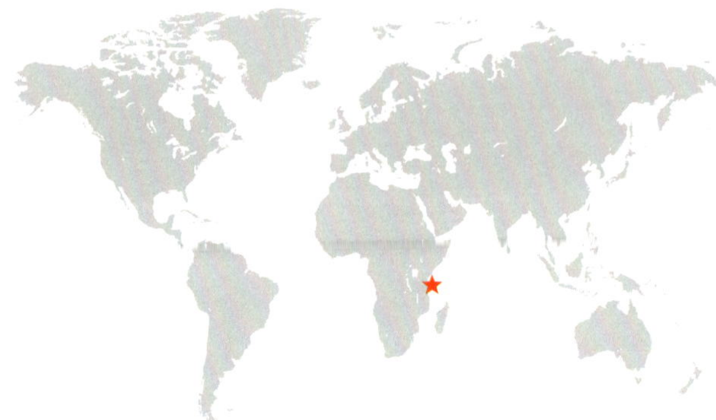

The Zones of Kilimanjaro
Tanzania

Stories about Mount Kilimanjaro often focus on its height and location. The tallest mountain in Africa is capped with snow and ice, despite sitting near the Equator. But it is also compelling for a different reason: To get to the icy summit, you must pass through incredibly diverse vegetation zones. The mountain rises from the hot, dry savanna, through rainforest and hardy scrublands, to a rocky and icy summit.

People have cultivated the lowlands ringing the mountain, which appear as patchy green areas. The continuous dark-green band is montane forest, which stretches from roughly 1,800 to 2,800 meters in elevation. The dark-green areas transition to a band of green-brown known as the moorland zone—colder, less humid, and full of short, hardy plants. The highest areas—the alpine desert and summit zones—are inhospitable to all but the most skilled mountain climbers.

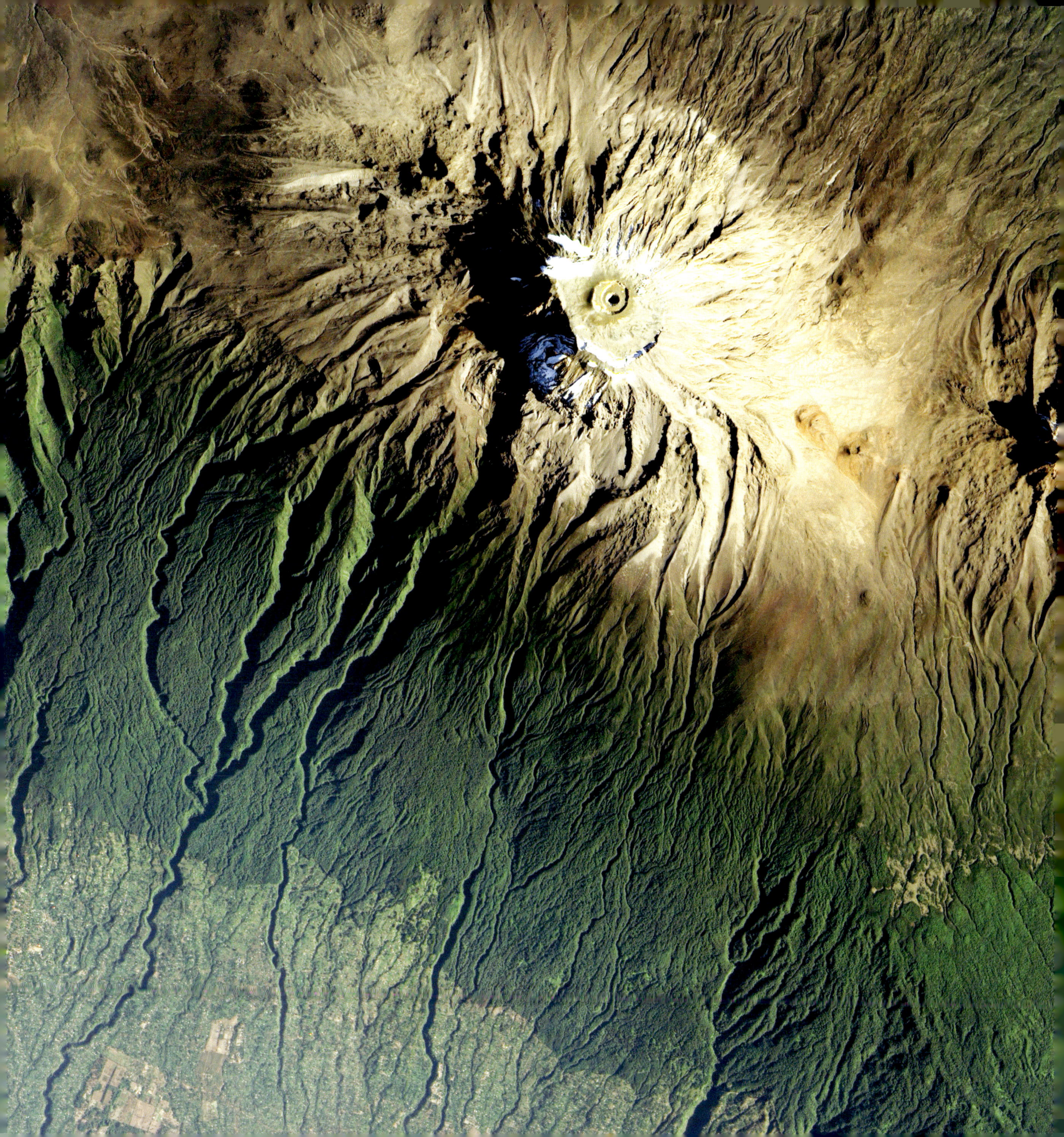

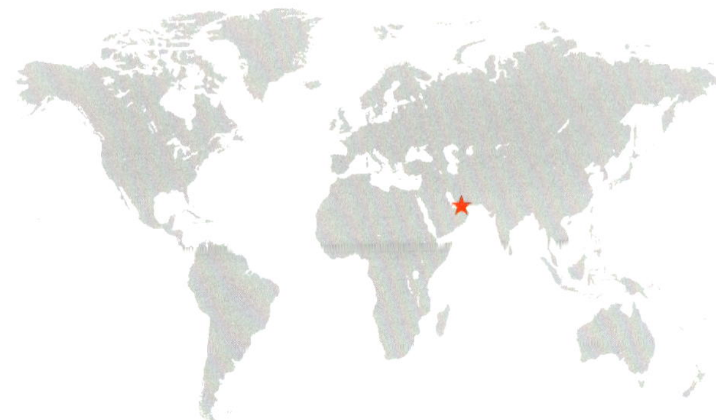

Liwa Oasis
United Arab Emirates

In the sandy tan terrain of the United Arab Emirates, on the northern edge of the Rub' al Khali, an oasis brings green to the desert. The T-shaped, 100-kilometer stretch of date plantations and small towns compose the Liwa Oasis, home to about 20,000 people in the emirate of Abu Dhabi. It is one of the largest oases on the Arabian Peninsula.

Bedouins tapped underground water supplies here at least five centuries ago, and date farms have proliferated. Drip irrigation and greenhouses now help conserve the precious water supply. Since rainfall is scarce in the region, much of the water comes from aquifers full of "fossil" water that accumulated more than 20,000 years ago and is now buried deep under the sand seas and limestone formations.

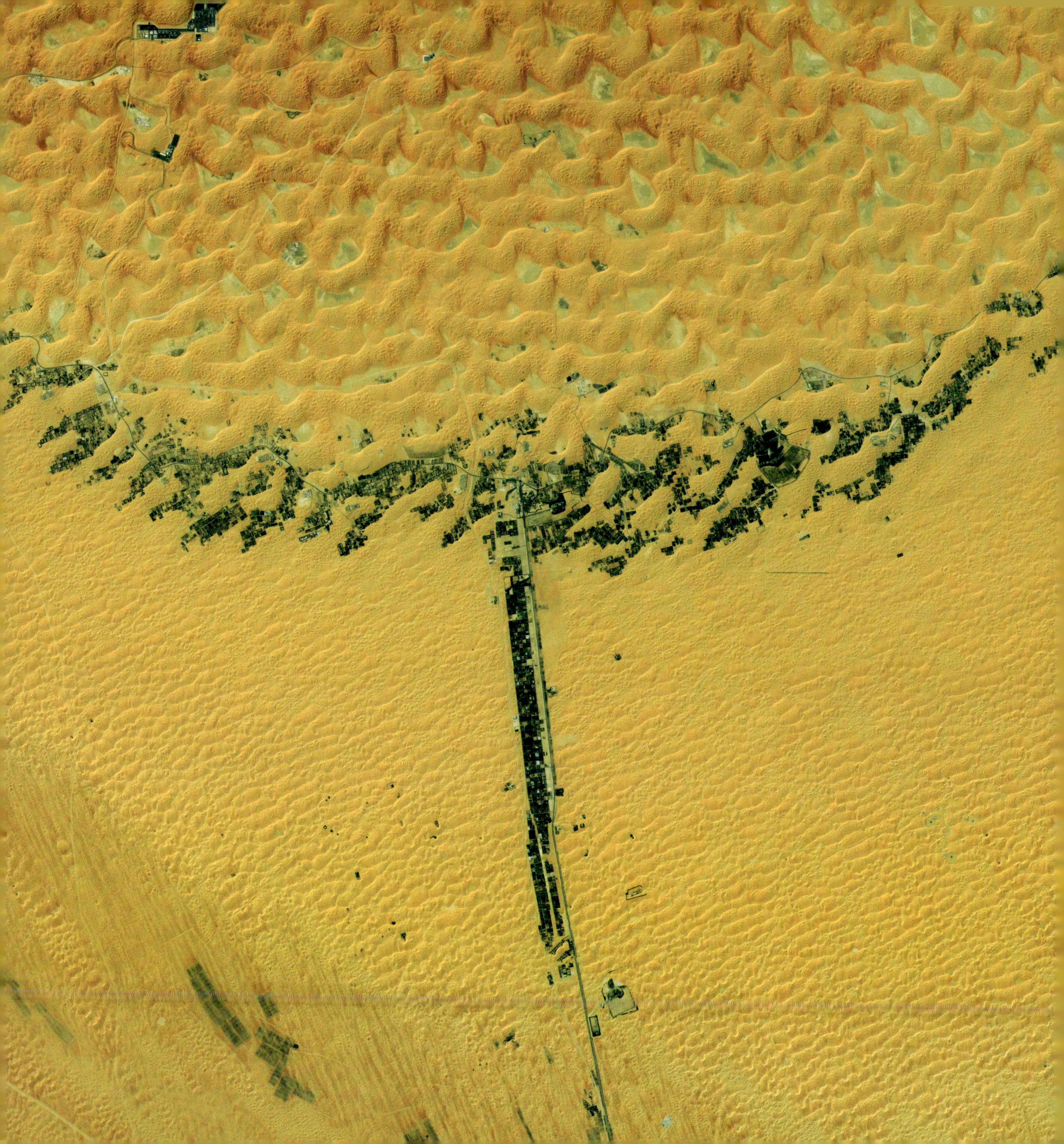

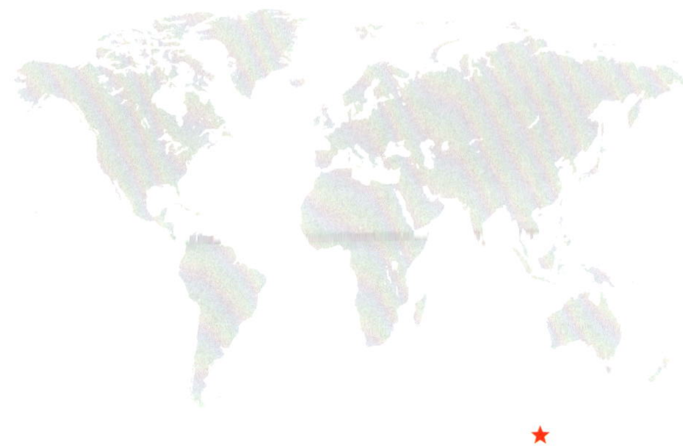

Don Juan Pond
Antarctica

In a valley in one of the most extreme environments on Earth lies the world's saltiest body of water. It rarely snows and never rains in the McMurdo Dry Valleys of Antarctica. Winter temperatures can drop to –50° Celsius, and the few ponds and lakes are capped by ice that is several meters thick.

Then there's Don Juan Pond. The ankle-deep pond in Upper Wright Valley is so salty that its calcium chloride–rich waters rarely freeze. With a salinity level of over 40 percent, Don Juan is significantly saltier than the Dead Sea and the Great Salt Lake.

The Earth Observing-1 satellite captured this image in January 2014. The ellipse-shaped lake is situated at the bottom of a basin between the Dais plateau and the Asgard Range to the south. It has a slightly darker hue than the salt-encrusted lake bottom around it.

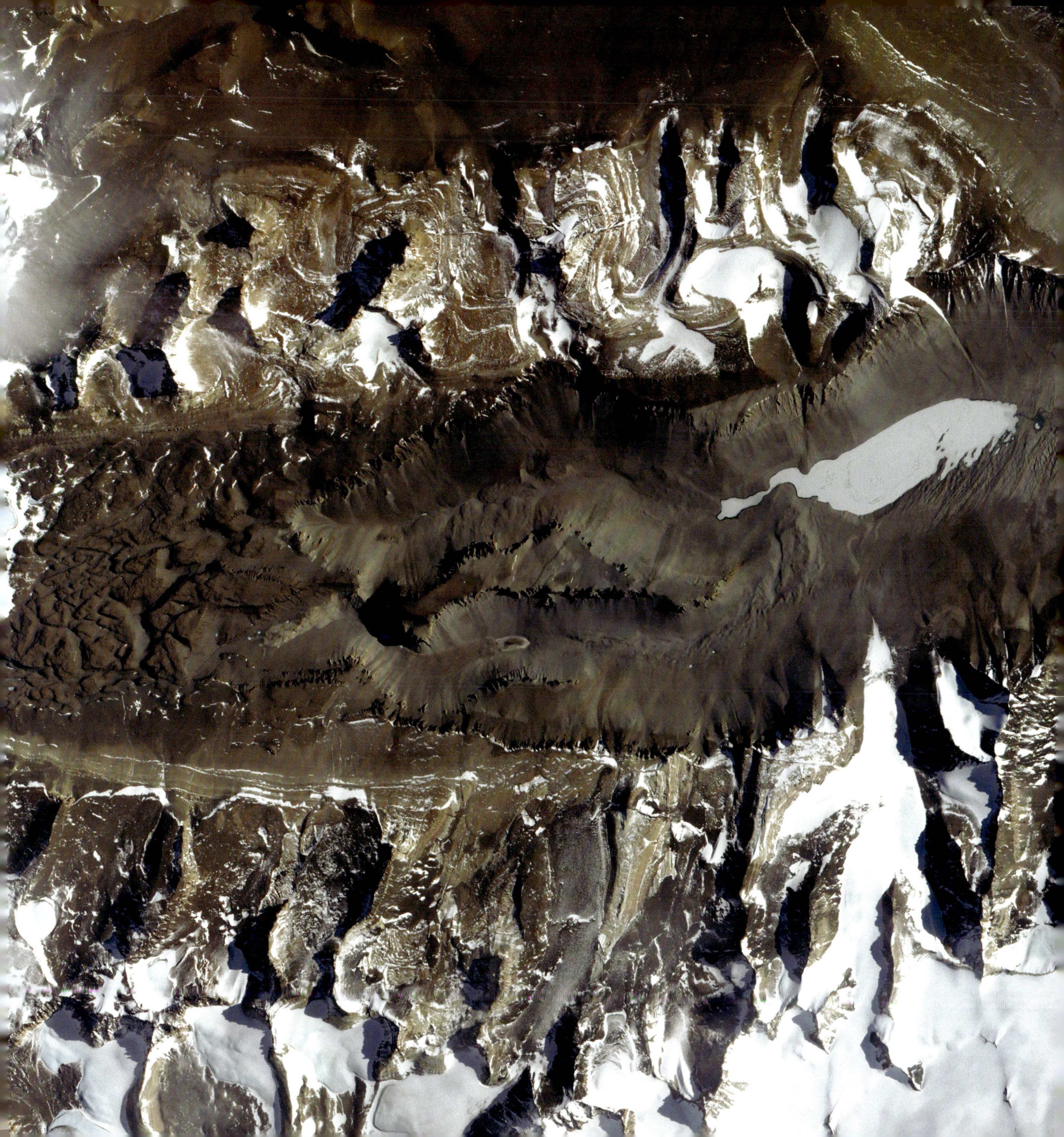

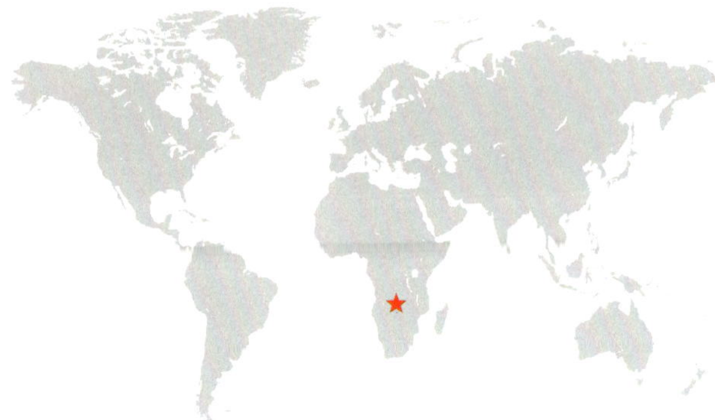

Linear Dunes, Caprivi Strip
Namibia

In far northeastern Namibia, there is a skinny stretch of land sandwiched between Angola, Botswana, and Zambia. The Caprivi Strip receives about 600 millimeters (24 inches) of rainfall each year. That's not a lot of rain—it tends to come in bursts that cause periodic floods—and it is a stark contrast to the much drier parts of the country.

Here the land is striped, as if a giant had dragged a rake over the landscape. Those stripes are linear dunes, and some of them are more than 100 kilometers (60 miles) long. Dunes generally form from wind-blown sand over many years, and one characteristic of linear dunes is that they tend to remain intact long after the dry conditions cease. And because they don't migrate like marching dunes, linear dunes preserve dirt and rocks that geologists can later use to understand past conditions.

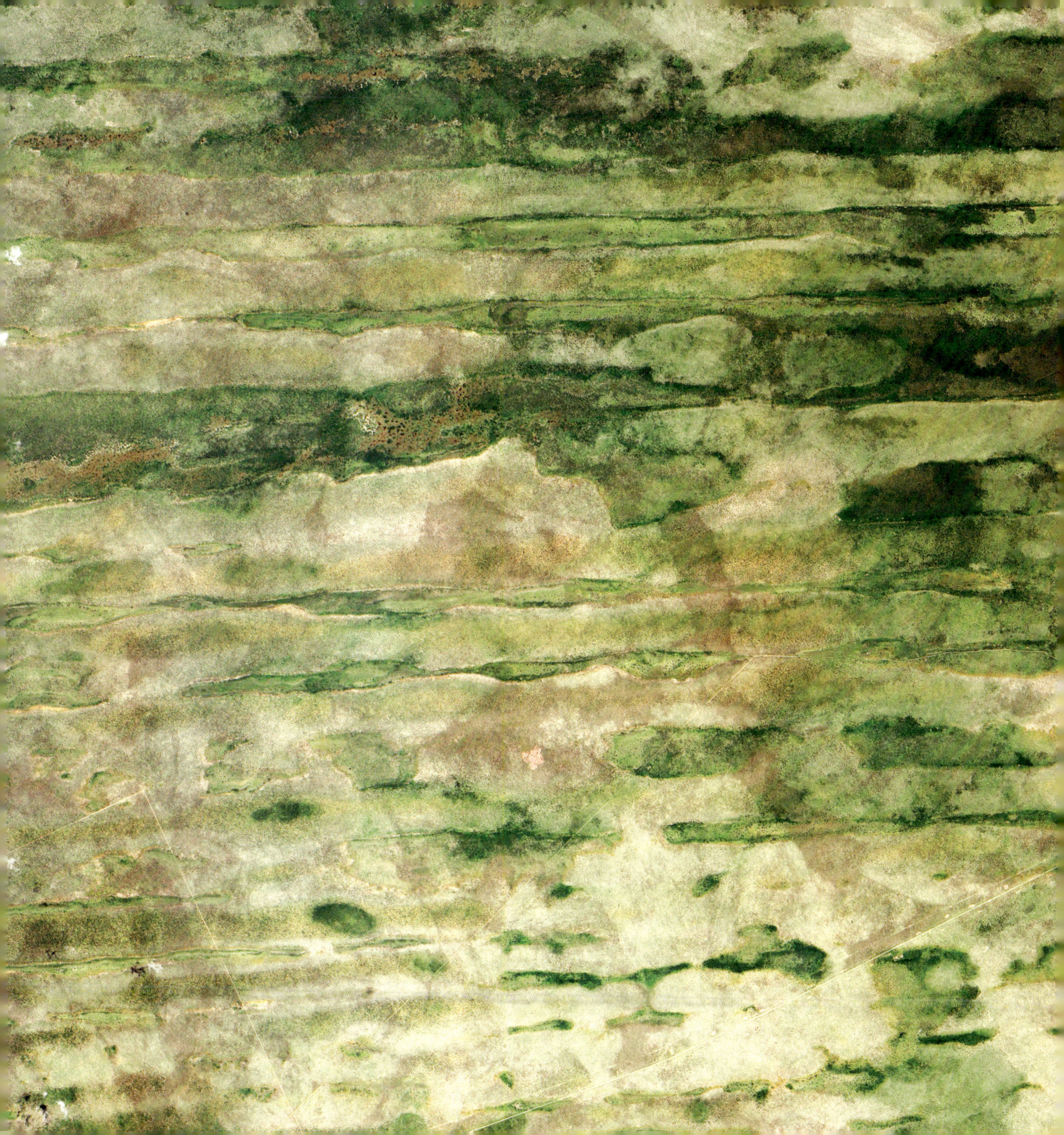

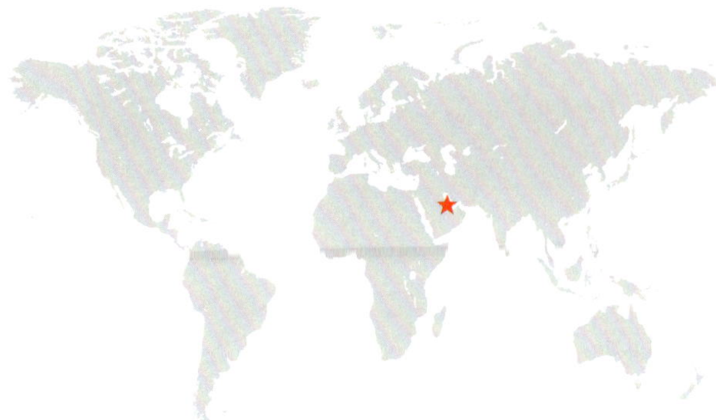

Harratt Lunayyir Lava Field
Saudi Arabia

In northwestern Saudi Arabia lies a field of volcanic lava. Known as Harratt Lunayyir, the lava field contains some 50 cones from volcanic eruptions over the past 10,000 years. The Terra satellite captured this false-color image in October 2006. Old lava flows appear as irregular, dark stains on an otherwise light-colored landscape. Like ink on an uneven surface, the lava has formed rivulets of rock that flow out in all directions.

Although one of the volcanic cones may have erupted as recently as the 10th century CE, scientists long believed the region to be geologically quiet until a seismic swarm and the opening of a crevice in 2009 suggested otherwise. The lava field Harratt Lunayyir lies about 200 kilometers (120 miles) from the tectonic spreading center under the Red Sea; magma can rise along the margins of such areas.

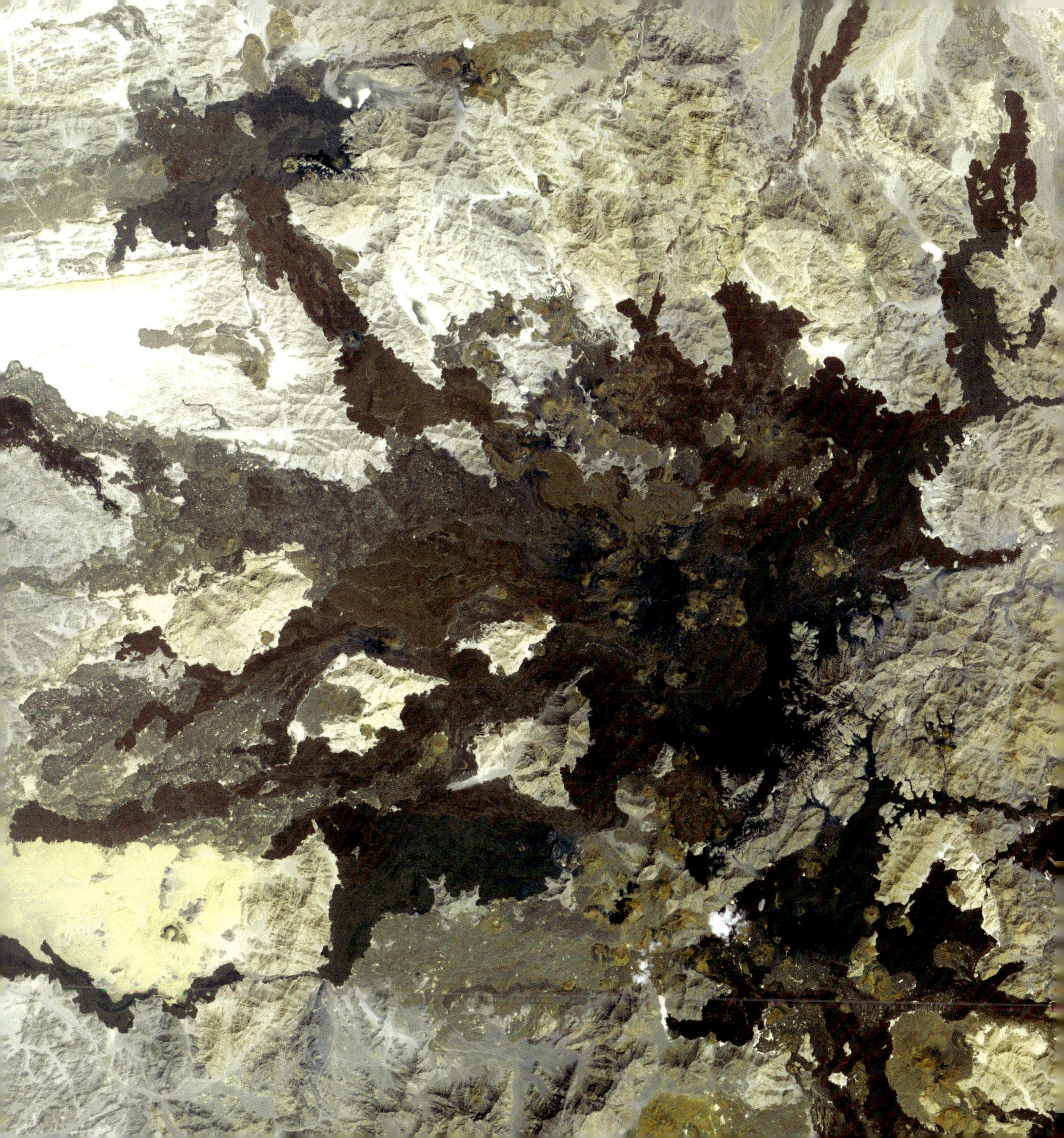

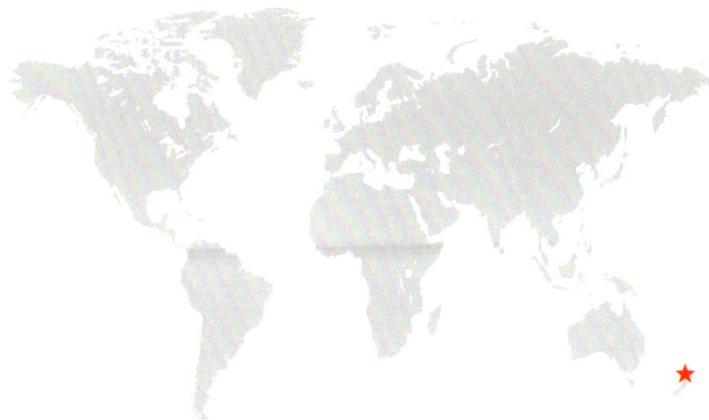

Taranaki and Egmont
New Zealand

The circular pattern of New Zealand's Egmont National Park stands out from space as a human fingerprint on the landscape. The park protects the forested and snow-capped slopes around Mount Taranaki (Mount Egmont to British settlers). It was established in 1900, when officials drew a radius of 10 kilometers around the volcanic peak. The colors differentiate the protected forest (dark green) from once-forested pasturelands (light- and brown-green).

Named by the native Maori people, Taranaki stands 2,518 meters (8,260 feet) tall, and it is one of the world's most symmetric volcanoes. It first became active about 135,000 years ago. By dating lava flows, geologists have figured out that small eruptions occur roughly every 90 years and major eruptions every 500 years. Landsat 8 acquired this image of Taranaki and the park in July 2014.

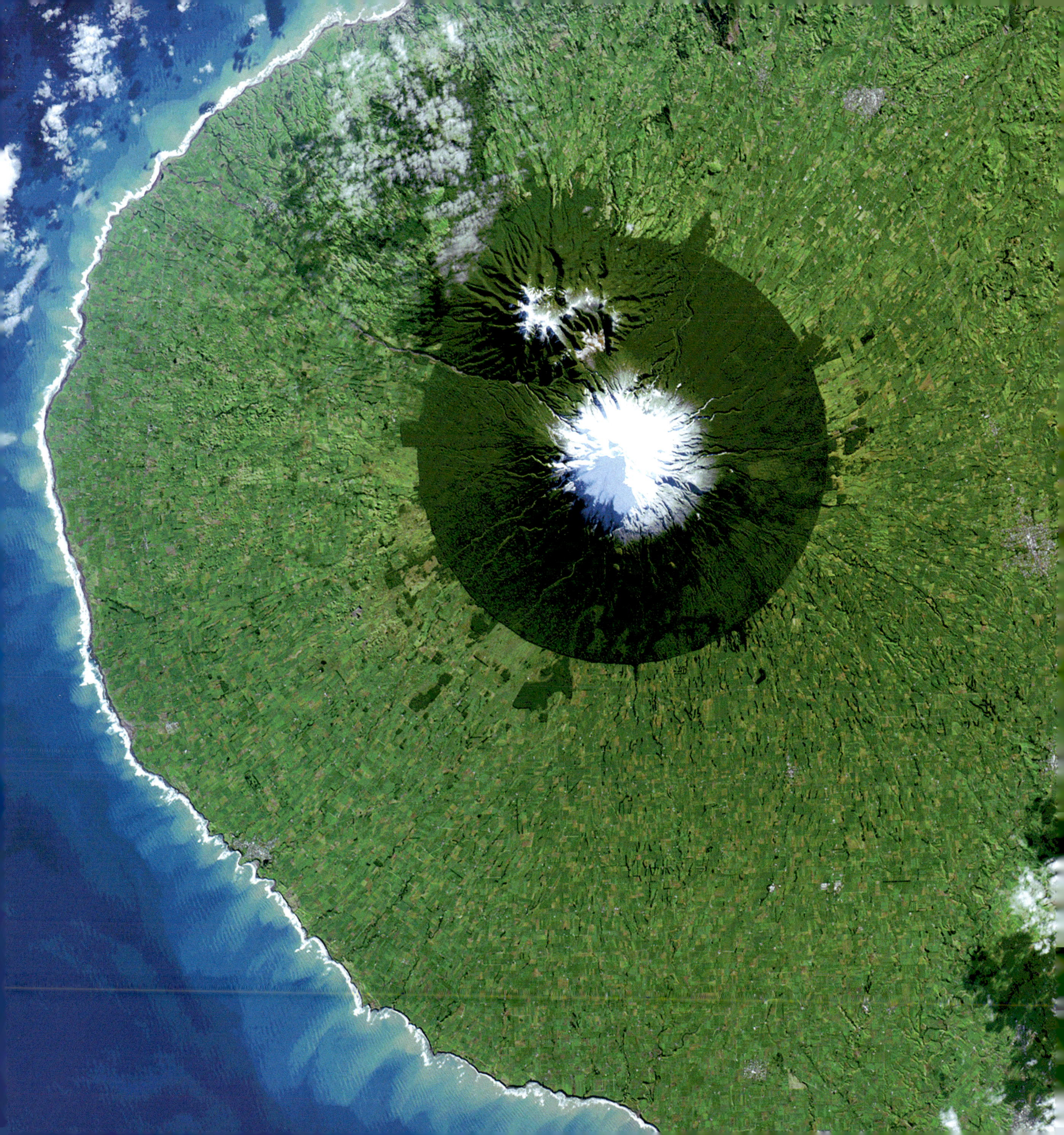

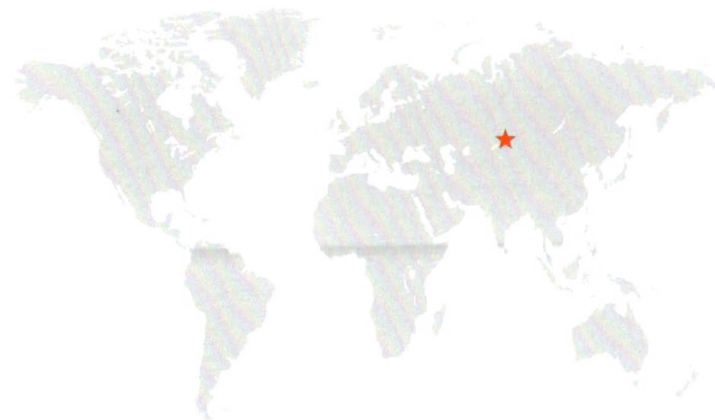

Cultivating a Border
China and Kazakhstan

While people often say borders are not visible from space, this line between eastern Kazakhstan and northwestern China could not be clearer. The border is made visible due to land-use policies.

With limited arable land and a large human population to feed, China farms just about any land that can be sustained for agriculture. In this Landsat 8 image from September 2013, fields are dark green in contrast to the surrounding dry landscape, a sign that the farms are irrigated.

While agriculture is important in the Kazakh economy, eastern Kazakhstan is a minor growing area for that country. A few rectangular shapes show that farming does occur. Much of the agriculture on the Kazakh side is rain-fed, so fields are tan like the surrounding, natural landscape.

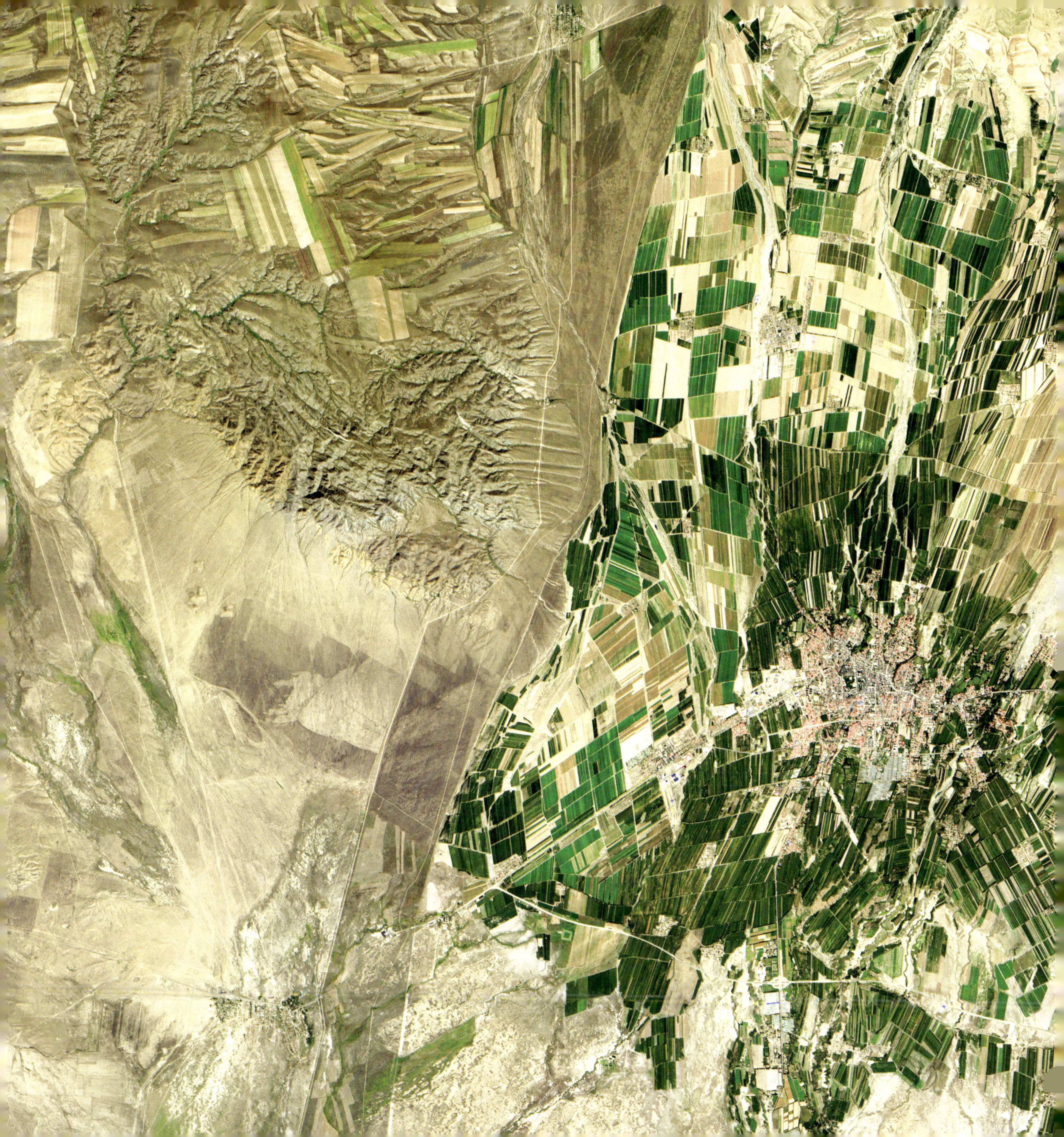

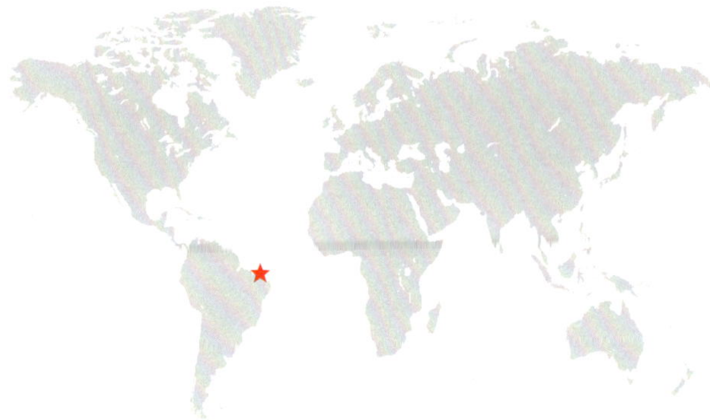

Barrier Islands
Brazil

Barrier islands are narrow strips of sand—often spits or sandbars that grow into full-blown, vegetated islands. They stretch from a few hundred meters to several kilometers wide. They run parallel to the coast, facing the sea, bearing the brunt of waves and wind, and protecting lagoons, bays, and coastal wetlands. And they move constantly, shaped and reshaped by currents, tides, people, and winds.

Barrier islands are found along the edge of every continent except Antarctica, and scientists and naturalists are still finding new ones. In June 2006, Landsat 5 captured this image of previously unrecognized barrier islands along the coast of Brazil between the Amazon River and São Luís. Brazil has the world's longest continuous chain of barrier islands—54 in total—extending more than 570 kilometers (350 miles) along the Atlantic coast.

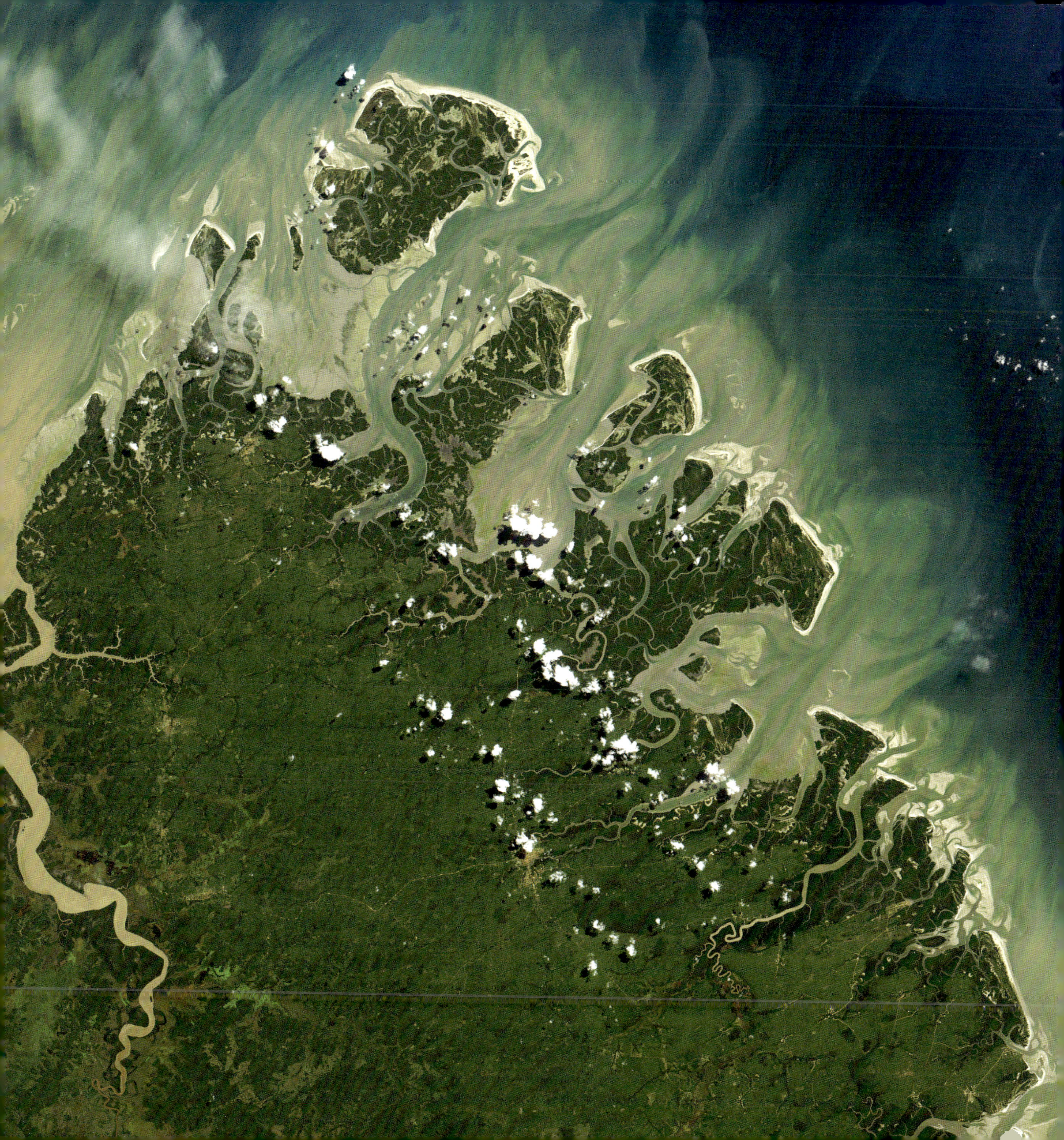

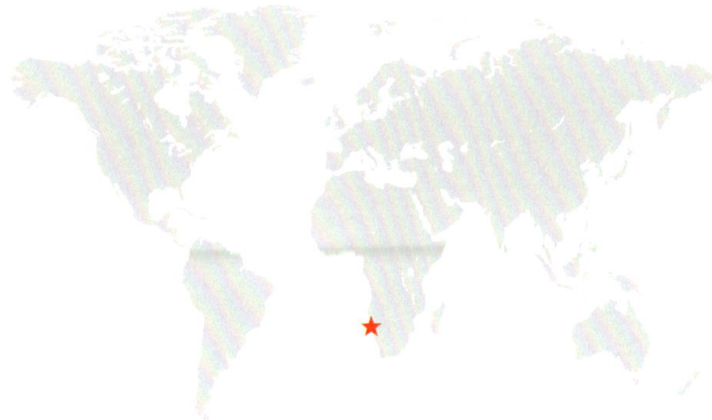

Tsauchab River Bed
Namibia

The Tsauchab River is a famous landmark for the people of Namibia and tourists. Yet few people have ever seen the river flowing with water. In times past, when the climate was more temperate, the Tsauchab likely reached the Atlantic coast, 55 kilometers to the west.

Like several other rivers around the Namib Desert, the Tsauchab brings sediment down from the hinterland toward the coastal lowland. This sediment is then blown from the river beds, and over tens of millions of years it has accumulated as the red dunes of the Namib Sand Sea.

In December 2009, an astronaut on the International Space Station caught this glimpse of the Tsauchab River bed jutting into the sea of red dunes. It ends in a series of light-colored, silty mud holes on the dry lake floor.

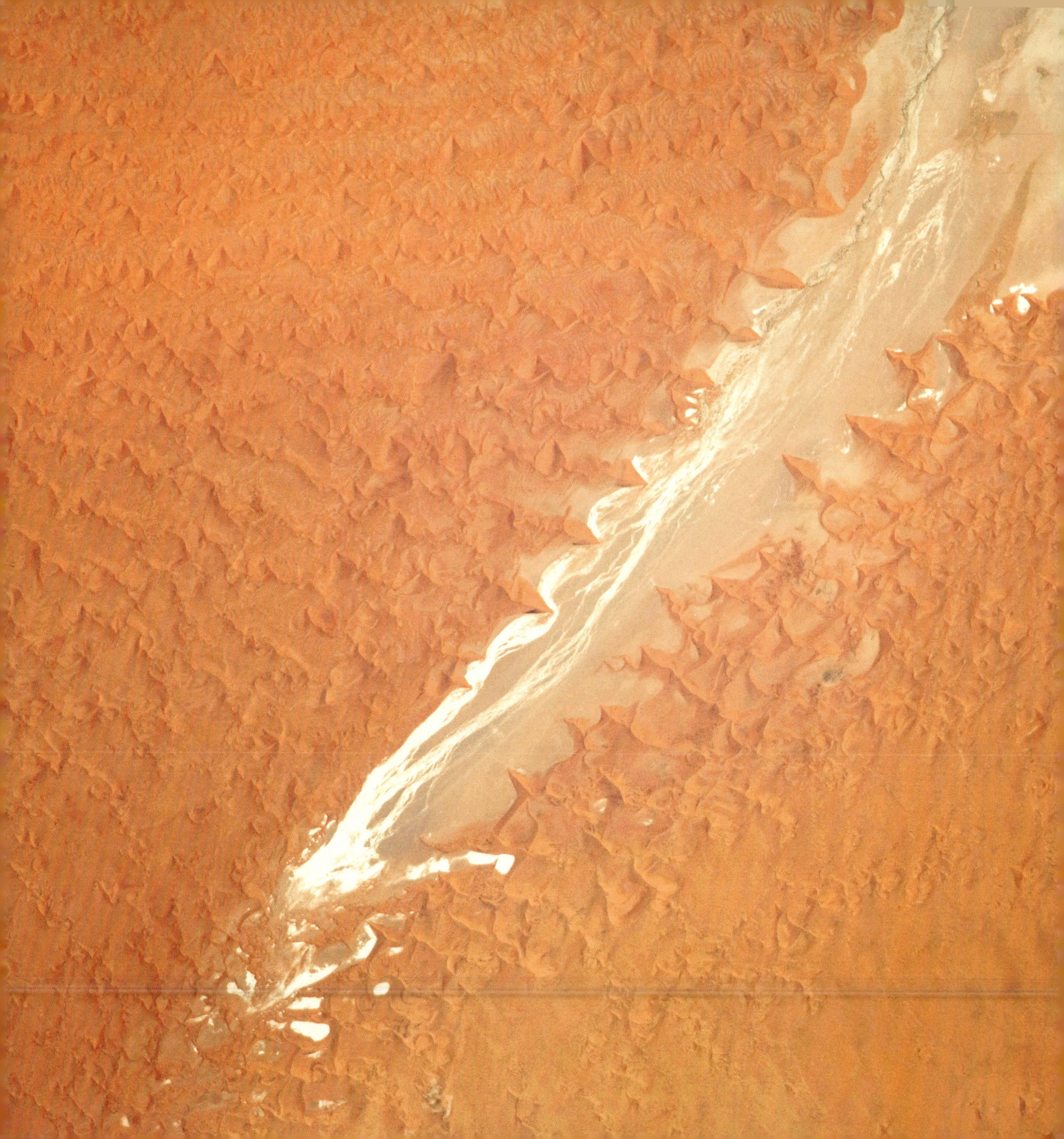

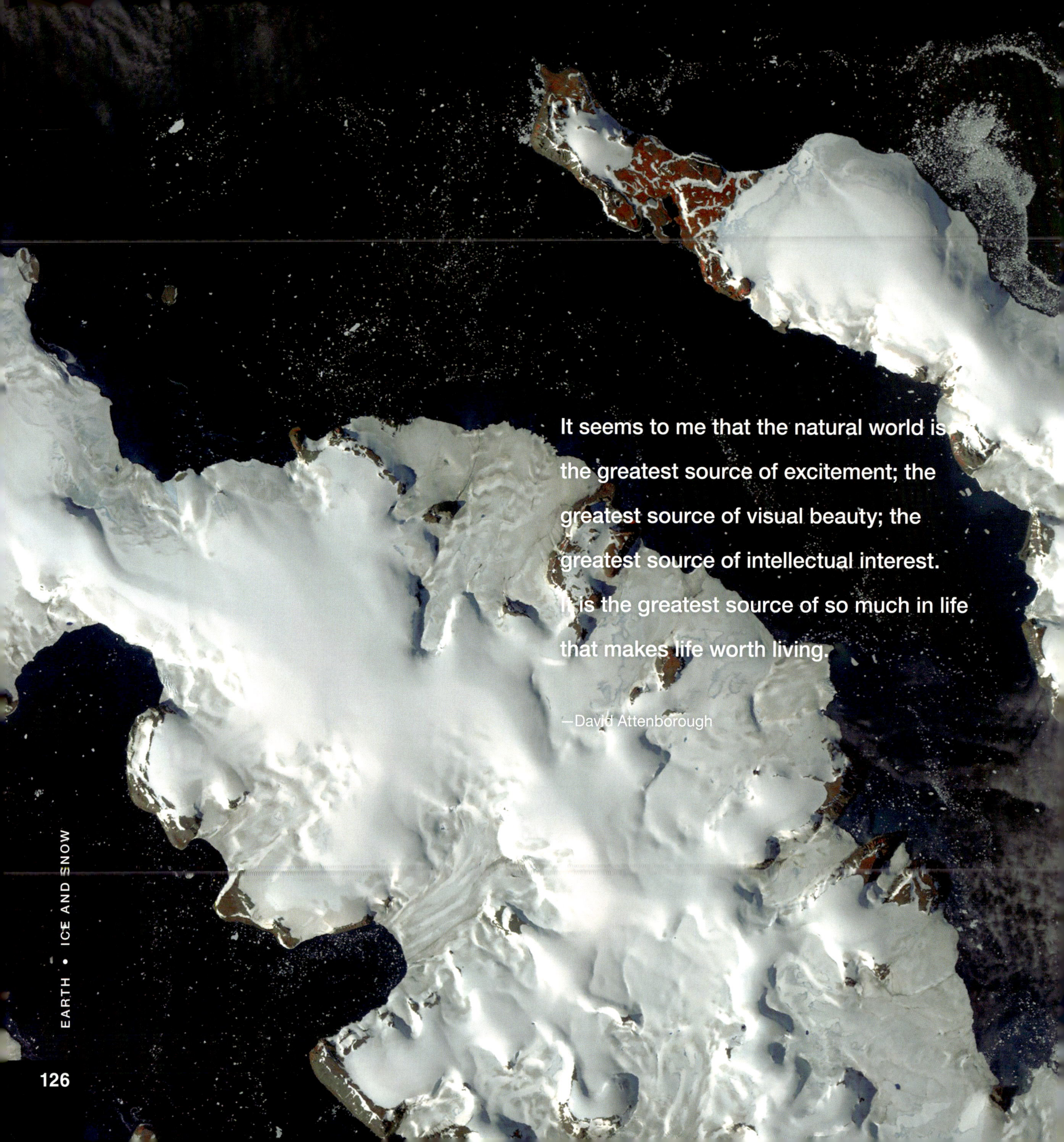

It seems to me that the natural world is the greatest source of excitement; the greatest source of visual beauty; the greatest source of intellectual interest. It is the greatest source of so much in life that makes life worth living.

—David Attenborough

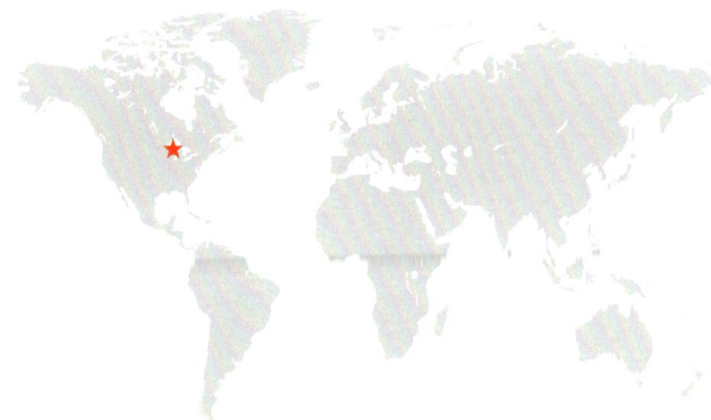

Swimming with Ice Cubes
United States

The ice season on the Great Lakes was longer in 2013–14 than anything in the satellite records to date or in anyone's memory. For nearly seven months, ice was afloat somewhere on the Great Lakes. In an average year, the lakes are ice-free by late April or early May—even as air temperatures onshore approached 27° Celsius (80° Fahrenheit) on some days. In 2014, the last ice melted in mid-June.

On May 23, 2014—the start of Memorial Day weekend (and unofficial start of summer in the United States)—Landsat 8 captured this image of ice in Lake Superior near Chequamegon Bay, Wisconsin. Satellite imagery of ice is one of many tools that government agencies use to manage shipping on the Great Lakes.

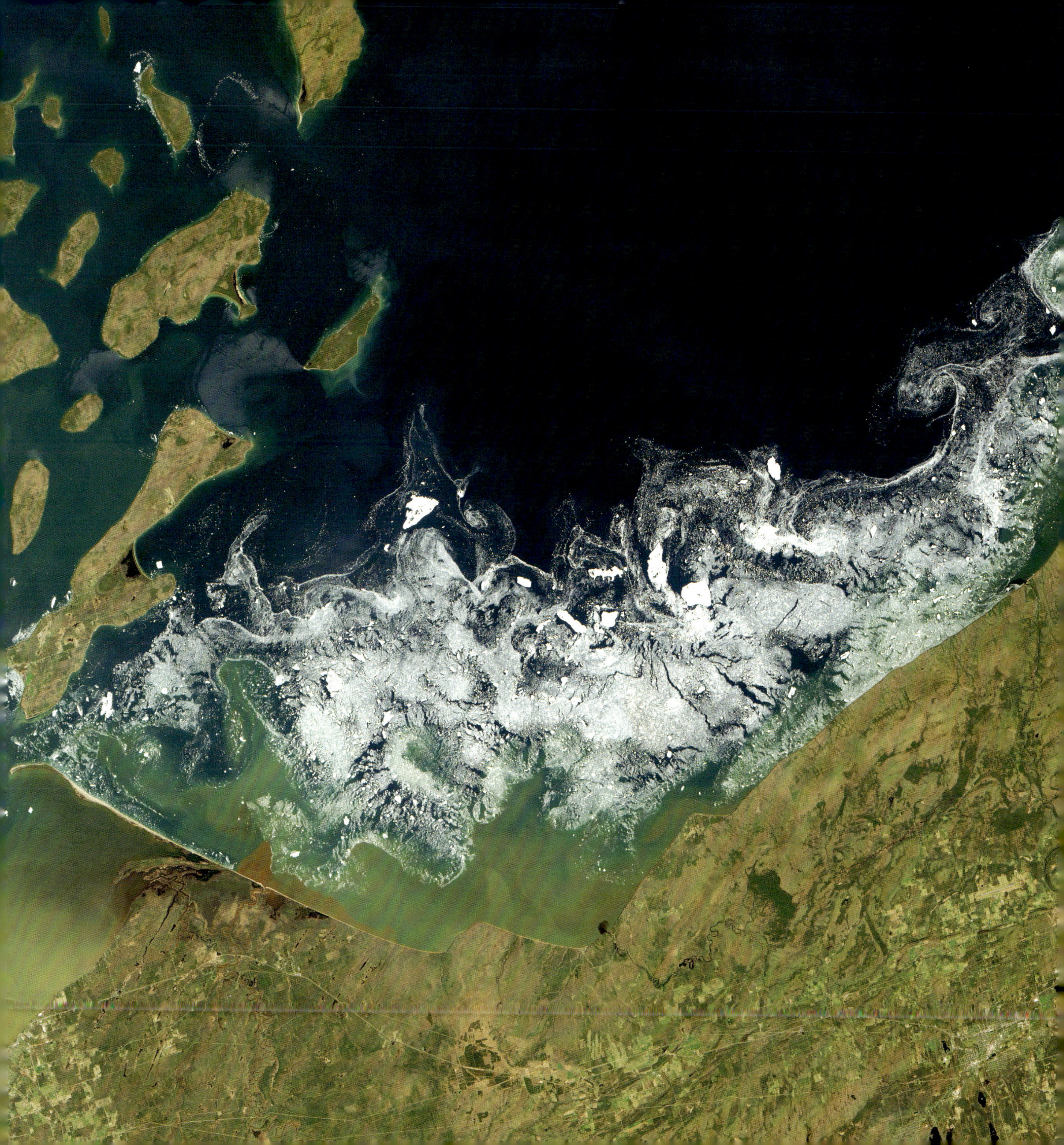

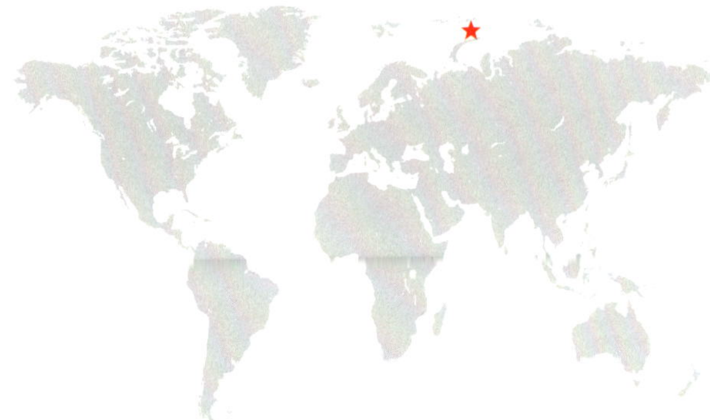

Franz Josef Land
Arctic Ocean

Located just 1,000 kilometers (600 miles) from the North Pole, Franz Josef Land is perpetually coated with ice. Glaciers cover roughly 85 percent of the archipelago's land mass, and sea ice floats in the channels between islands even in the summertime. The Terra satellite observed some of the islands in visible and near-infrared light in August 2011.

The amount of sea ice filling the channels between the islands of Franz Josef Land varies from summer to summer. Most of the ice in this scene is anchored to land, as large glaciers blanket the islands. Yet today's glaciers are tiny compared to the ice sheet that dominated the region about 20,000 years ago. Raised beaches, which preserve evidence of land rising as the crushing weight of overlying glaciers eases (known as isostatic rebound), were first recognized on the islands in the late 19th century.

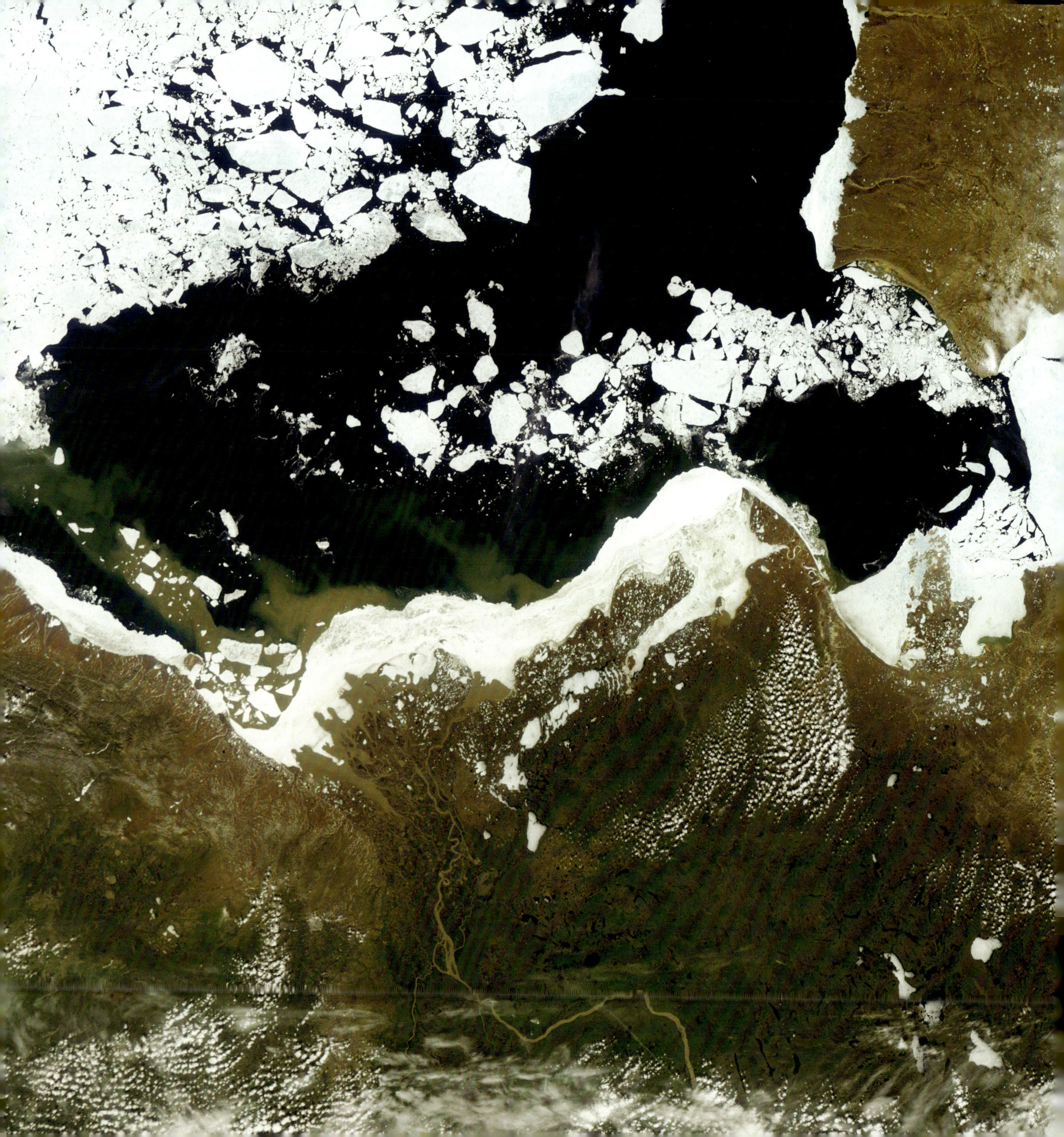

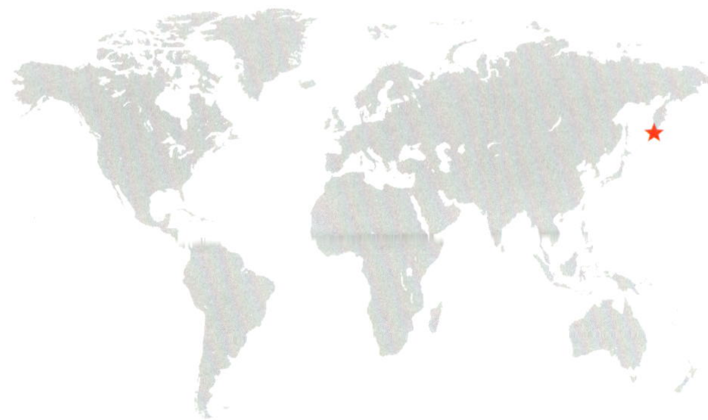

Sea Ice at Shikotan
Japan and Russia

Ostrov Shikotan is a volcanic island at the southern end of the Kuril chain. At about 43 degrees north—more than halfway to the Equator—Shikotan lies along the extreme southern edge of winter sea ice in the Northern Hemisphere. The Earth Observing-1 satellite captured this image of swirling blue-gray sea ice around Shikotan in February 2011.

The ice here tends to move with currents and eddies, which has shaped it into rough circles. The eddies may result from opposing winds from the north and southwest.

Uneven snow cover exaggerates the island's rugged appearance. Multiple forces have shaped Shikotan over millions of years. It has been battered by tsunamis—although wind, rain, and tectonic forces likely play a greater role in shaping the surface.

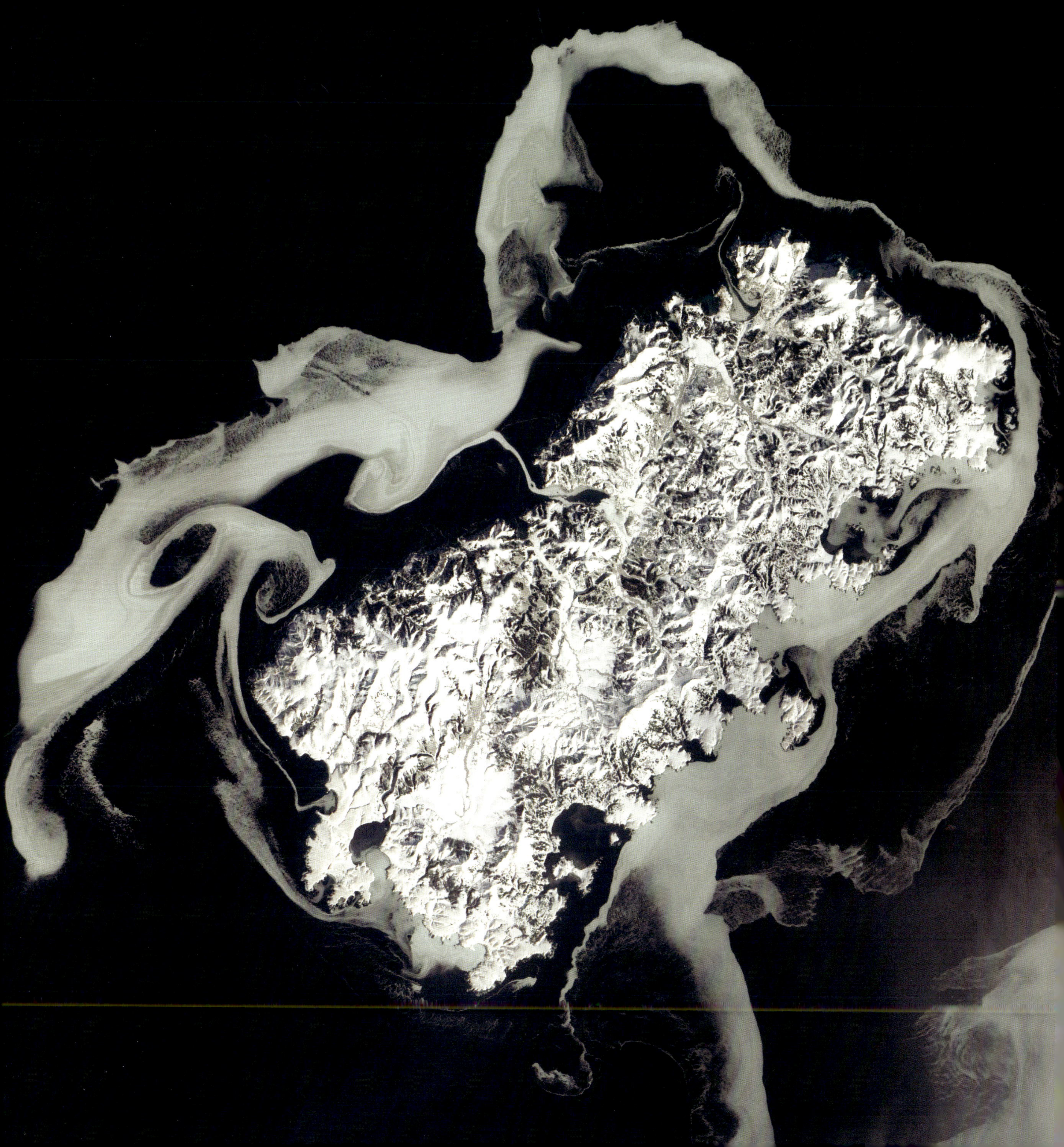

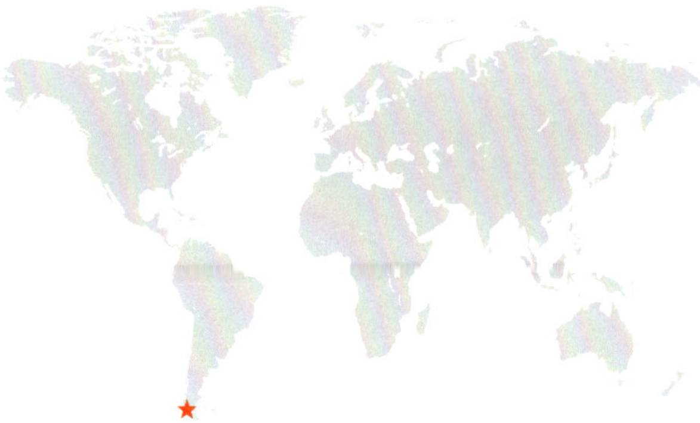

North Patagonian Icefield
South America

Forests, grasslands, deserts, and mountains are all part of the Patagonian landscape that spans more than a million square kilometers of South America. Toward the western side, expanses of dense, compacted ice stretch for hundreds of kilometers of the Andes mountain range in Chile and Argentina. The two lobes of the Patagonian icefields—north and south—are what is left of a much more expansive ice sheet that reached its maximum size about 18,000 years ago. The modern icefields are just a fraction of their previous size, though they remain the southern hemisphere's largest expanse of ice outside of Antarctica.

The northern icefield covers about 4,000 square kilometers and has 30 significant glaciers along its perimeter. In April 2017, Landsat 8 captured this rare cloud-free view of a portion of the icefield.

Ice creeps downslope through mountain valleys and exits through so-called "outlet glaciers." Many come to an abrupt end on land, while others terminate in water. The San Rafael and San Quintín glaciers (shown at the right) are the icefield's largest. Both have been receding rapidly in the past 30 years.

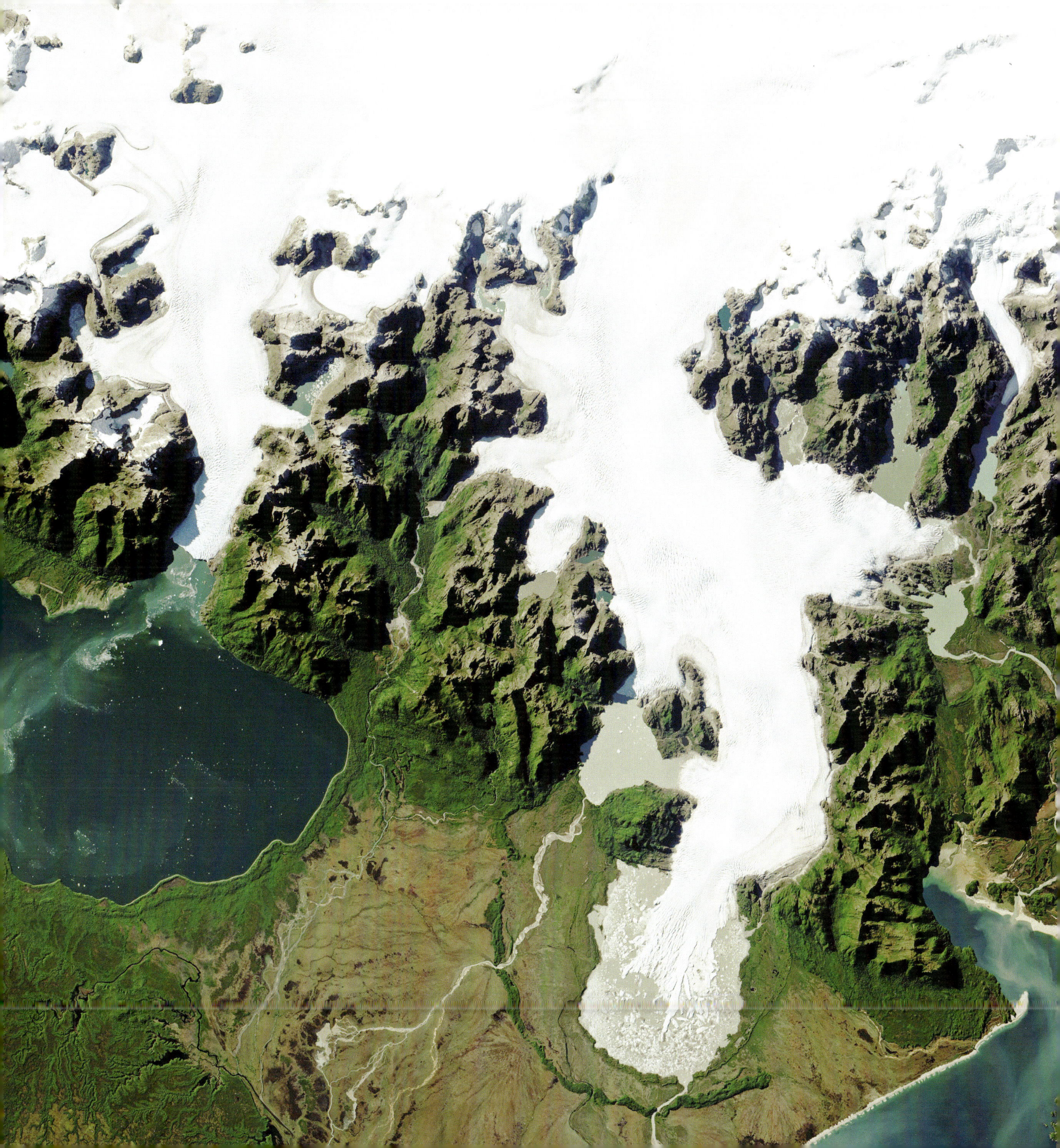

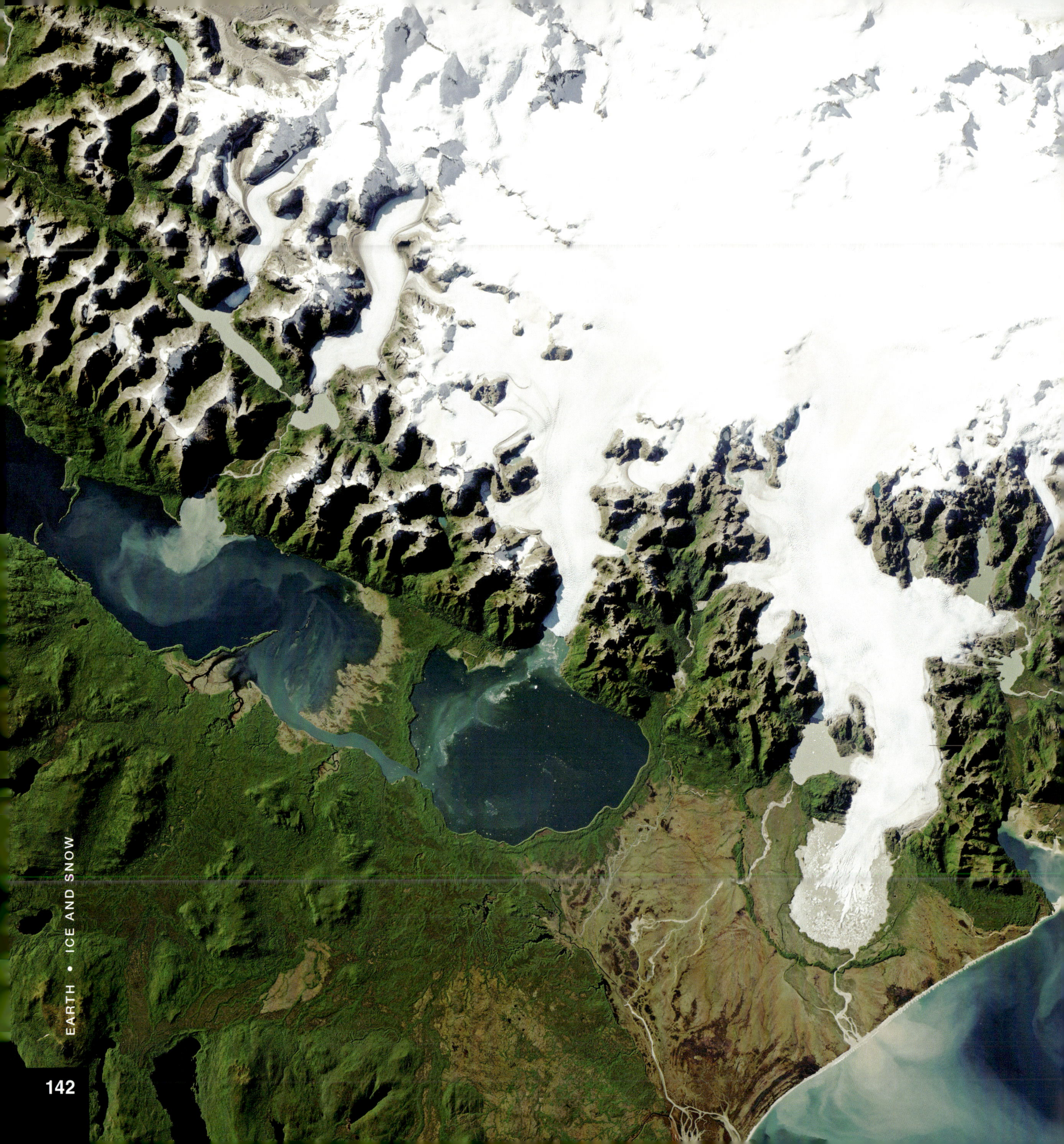

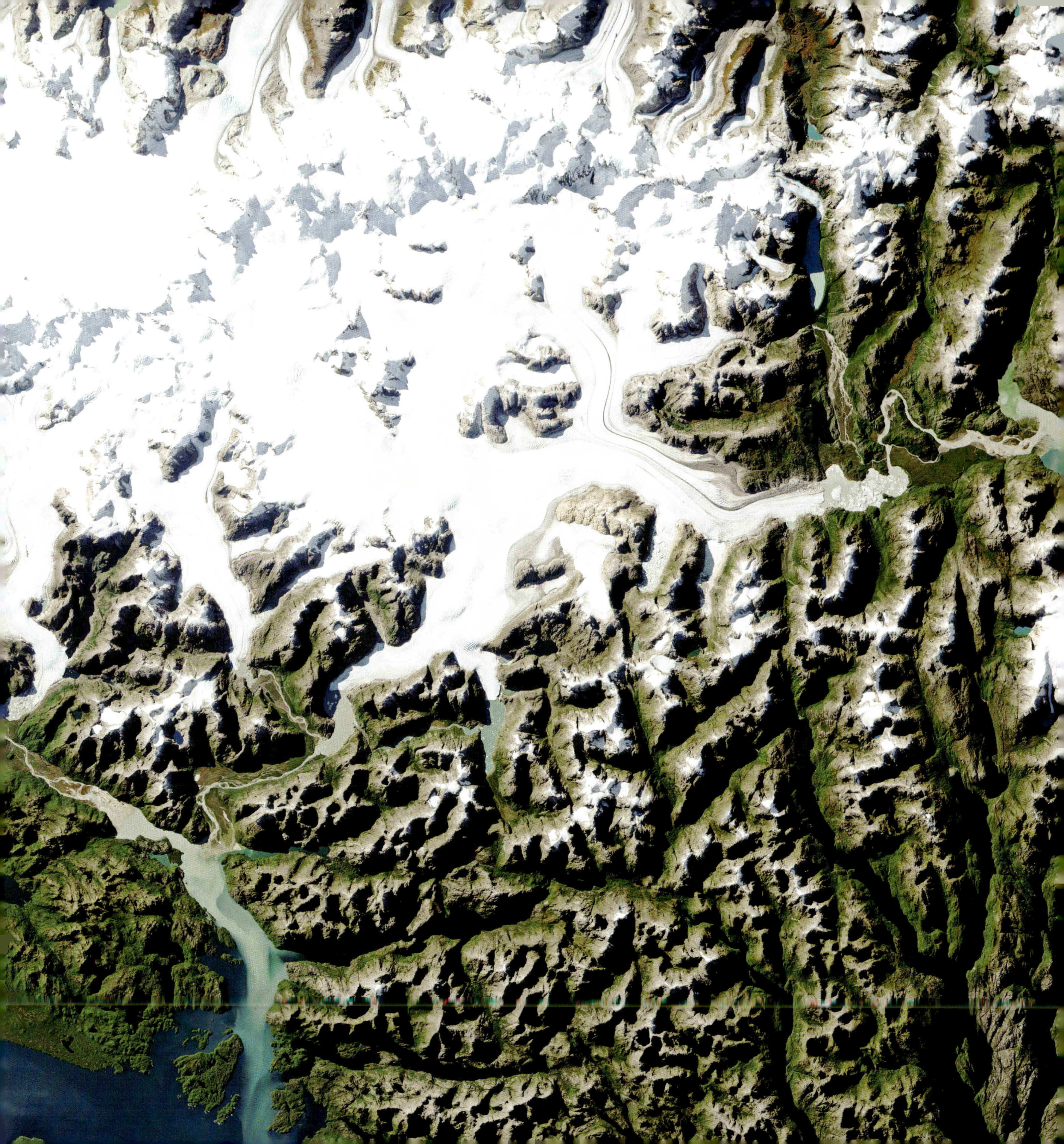

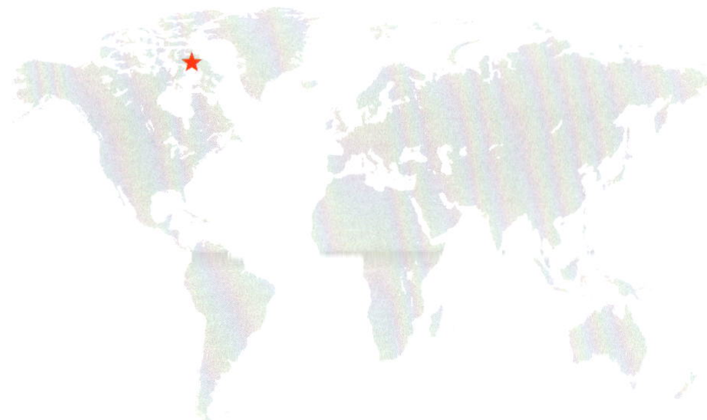

Manning Island and Foxe Basin
Canada

Although it may look like a microscope's view of a thin slice of mineral-speckled rock, this image was actually acquired in space by the Earth Observing-1 satellite in July 2012. It shows a small set of islands and a rich mixture of ice in Foxe Basin, the shallow northern reaches of Hudson Bay.

The small and diverse sizes of the ice floes indicate that they were melting. The darkest colors in the image are open water. Snow-free ice appears gray, while snow-covered ice appears white. The small, dark features on many of the floes are likely melt ponds.

Foxe Basin sea ice is known for having an unusual brown color due to staining from sediment from rivers. Also, the basin is shallow enough that it is often rich with marine sediments kicked up from the bay floor. The Manning Islands stand to the lower left, beneath the bright white wedge.

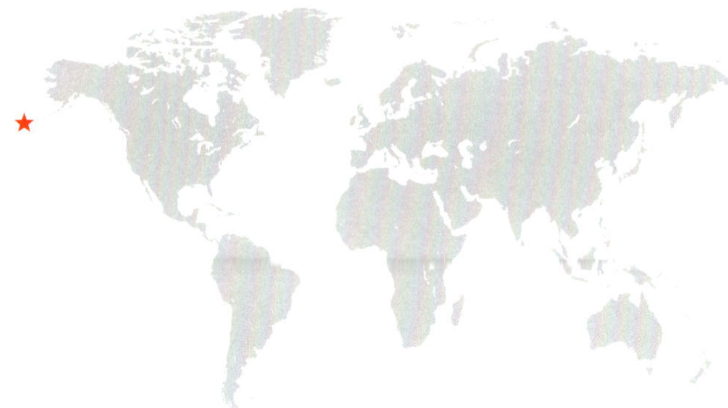

Ice Water
United States

As temperatures rise in the summer, turquoise splotches of color begin to speckle the icy surfaces of the Arctic. Those splashes of blue are melt ponds—areas where snow has melted and pooled in low spots atop glaciers and sea ice. During an airborne research campaign in July 2014, a scientist shot this photograph while flying over a glacier in southeastern Alaska. Chunks of ice float on the pond's turquoise water.

Many questions remain about the impact of melt ponds on the Arctic. Compared to bright white snow and ice, liquid water absorbs much more heat from sunlight. So when a pool of water forms on top of ice, it changes the heat balance. The water warms in the sunlight and can speed the melting of surrounding ice, influencing the overall melting and movement of ice sheets and sea ice.

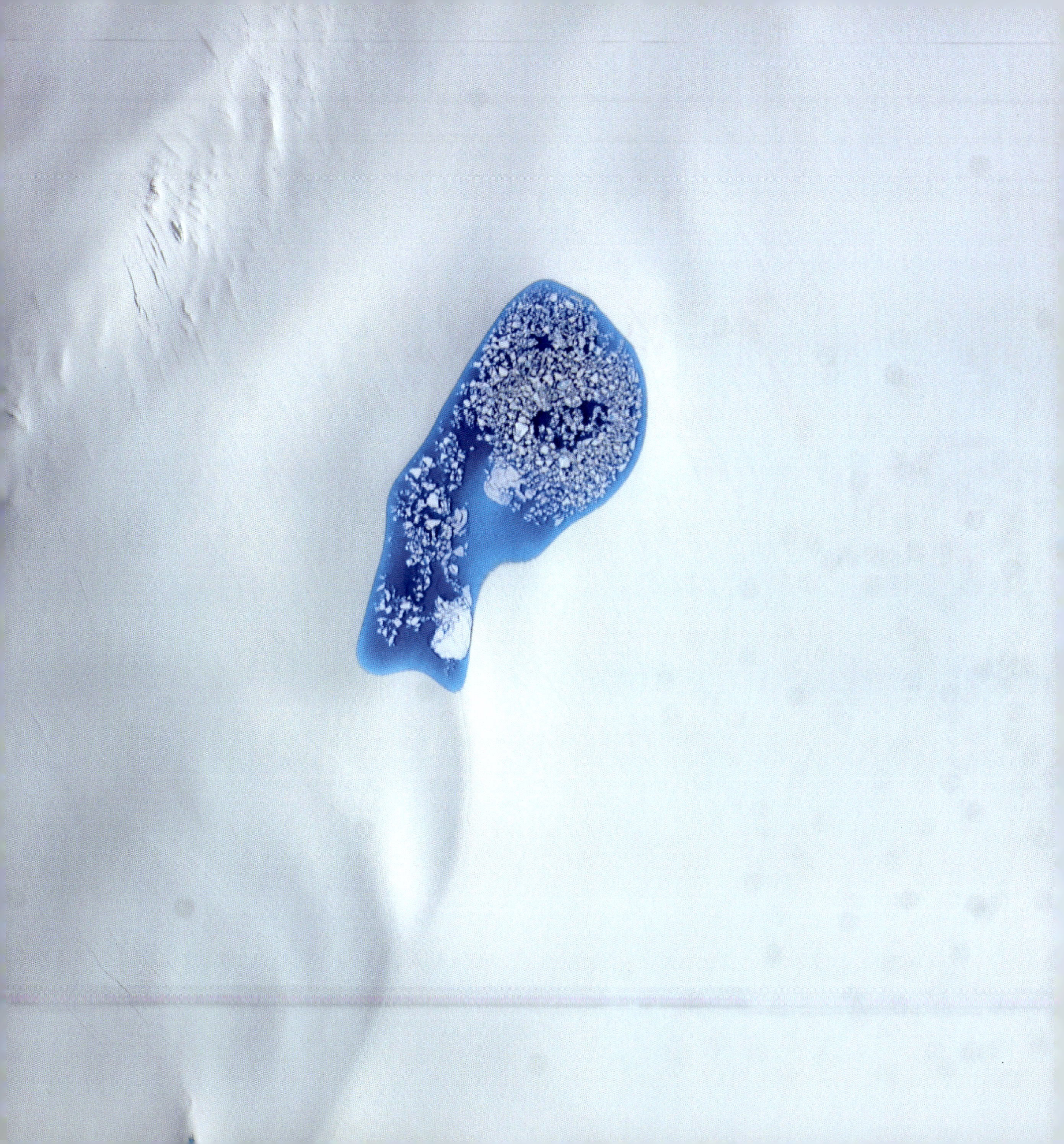

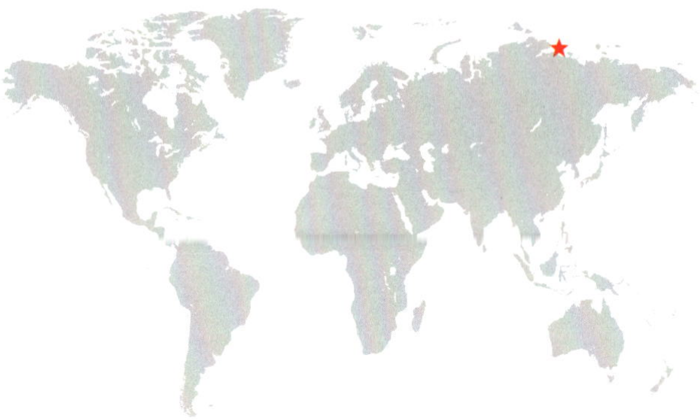

Omulyakhskaya and Khromskaya Russia

Along the northern Siberian coast, near Omulyakhskaya and Khromskaya Bays, the landscape is dotted with lakes. Known as thermokarst lakes, these pools are made from the thawing of frozen soil, or permafrost, and the accumulation of that melt water in low spots in the terrain.

Although far too cold for a swim, the water is generally warm compared to the surrounding soil, so it can slowly thaw more permafrost and make the lake deepen and expand over time. Occasionally the basins merge or even drain into streams and the bay. Dark brown spots in the image are probably locations of former thermokarst lakes.

Because thawing permafrost and thermokarst lakes release carbon and methane—both greenhouse gases—scientists monitor these landscapes closely because of their implications for future climate.

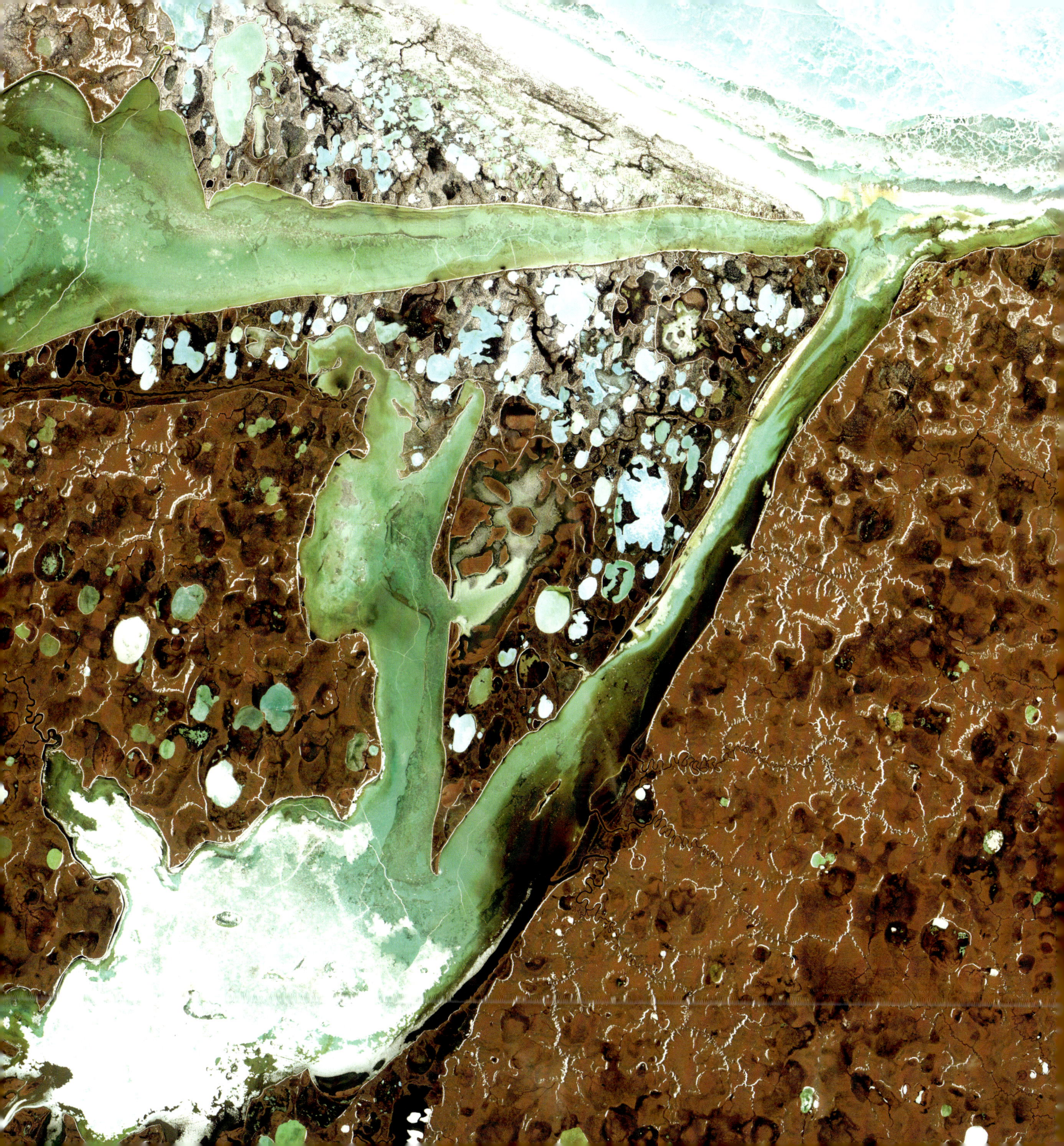

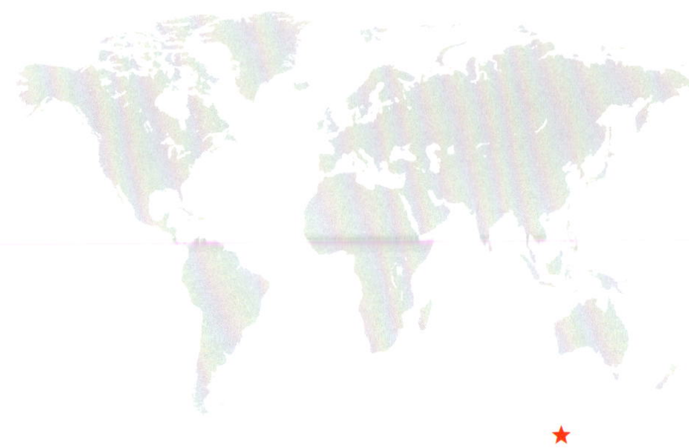

Phytoplankton on Ice
Antarctica

It may look like someone dyed the water, but the green hue visible off the coast of Antarctica is entirely natural. Granite Harbor, a cove near Antarctica's Ross Sea, got its color from phytoplankton at the water's surface. These microscopic, plant-like organisms typically flourish here in spring and summer, when the edge of the sea ice recedes and there is ample sunlight. But scientists have noticed that, given the right conditions, they can grow in autumn, too. In March 2017, Landsat 8 captured such an event in this image.

Sea ice, winds, sunlight, nutrient availability, and predators all factor into whether plankton can grow in large enough quantities to color the slush-ice and make it visible from space. Phytoplankton are important for the ecology of the Southern Ocean, as they are an abundant food source for zooplankton, fish, and other marine species.

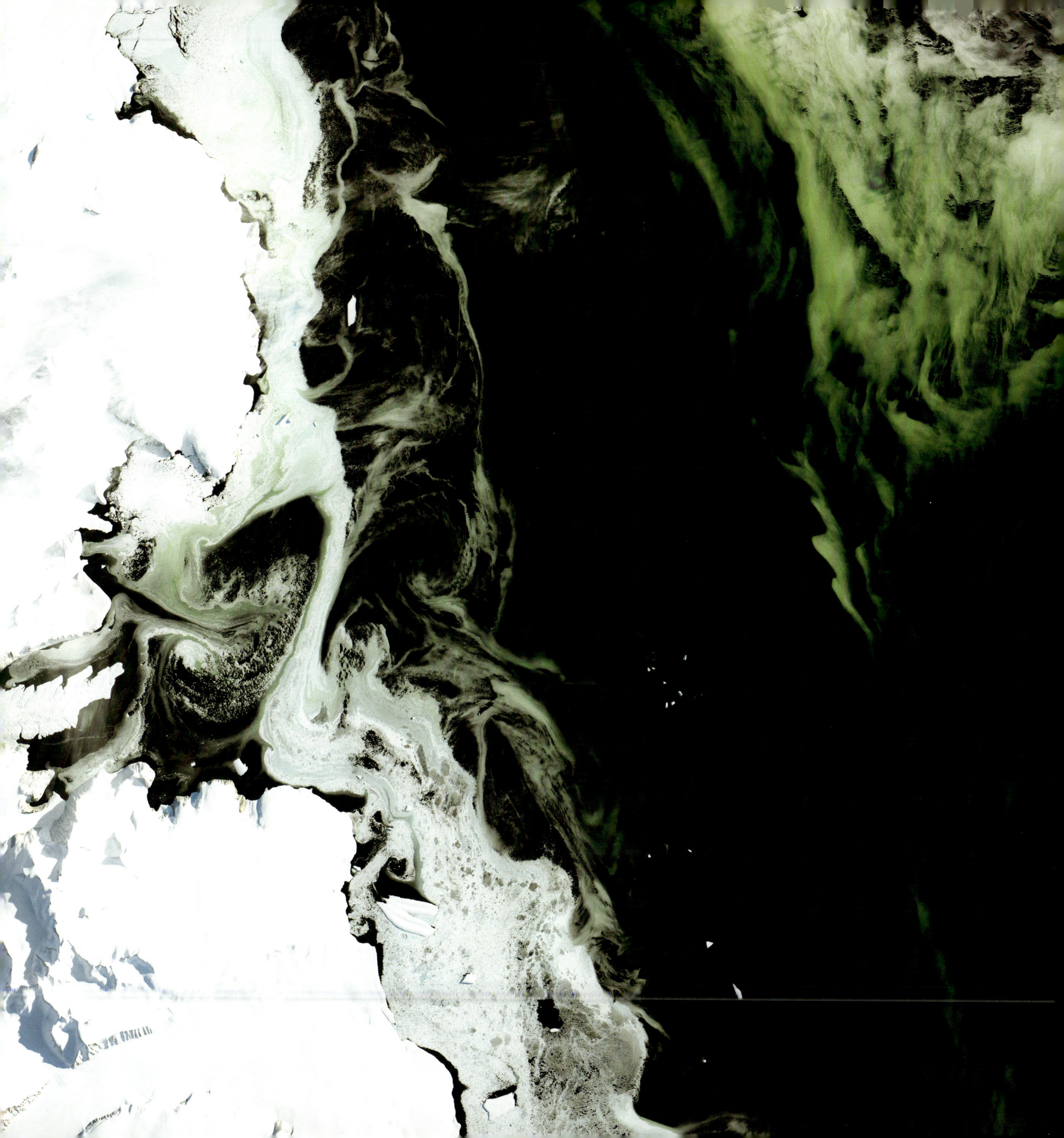

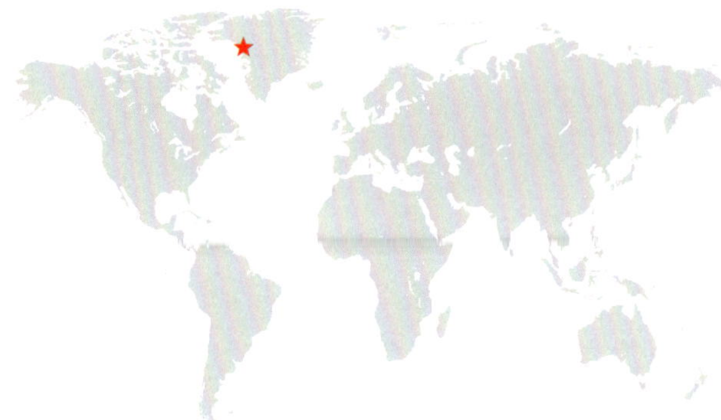

Heart-Shaped Uummannaq
Greenland

It is no mystery how Uummannaq Island got its name. In Greenlandic, the word means "heart-shaped," an apt description for the multi-peaked mountain that towers over the island.

Located off the coast of northwestern Greenland, the mountain's granite and gneiss peak rises sharply from sea level to about 1,170 meters (3,840 feet). The rock that makes up Uummannaq is ancient, likely forming 3.0 to 2.8 billion years ago.

Well north of the Arctic Circle, Uummannaq Island is home to one of the most northerly towns in Greenland. The Earth Observing-1 satellite captured this image in May 2012. Sea ice still surrounded the island, but breaks in the ice—called leads—exposed seawater beneath it.

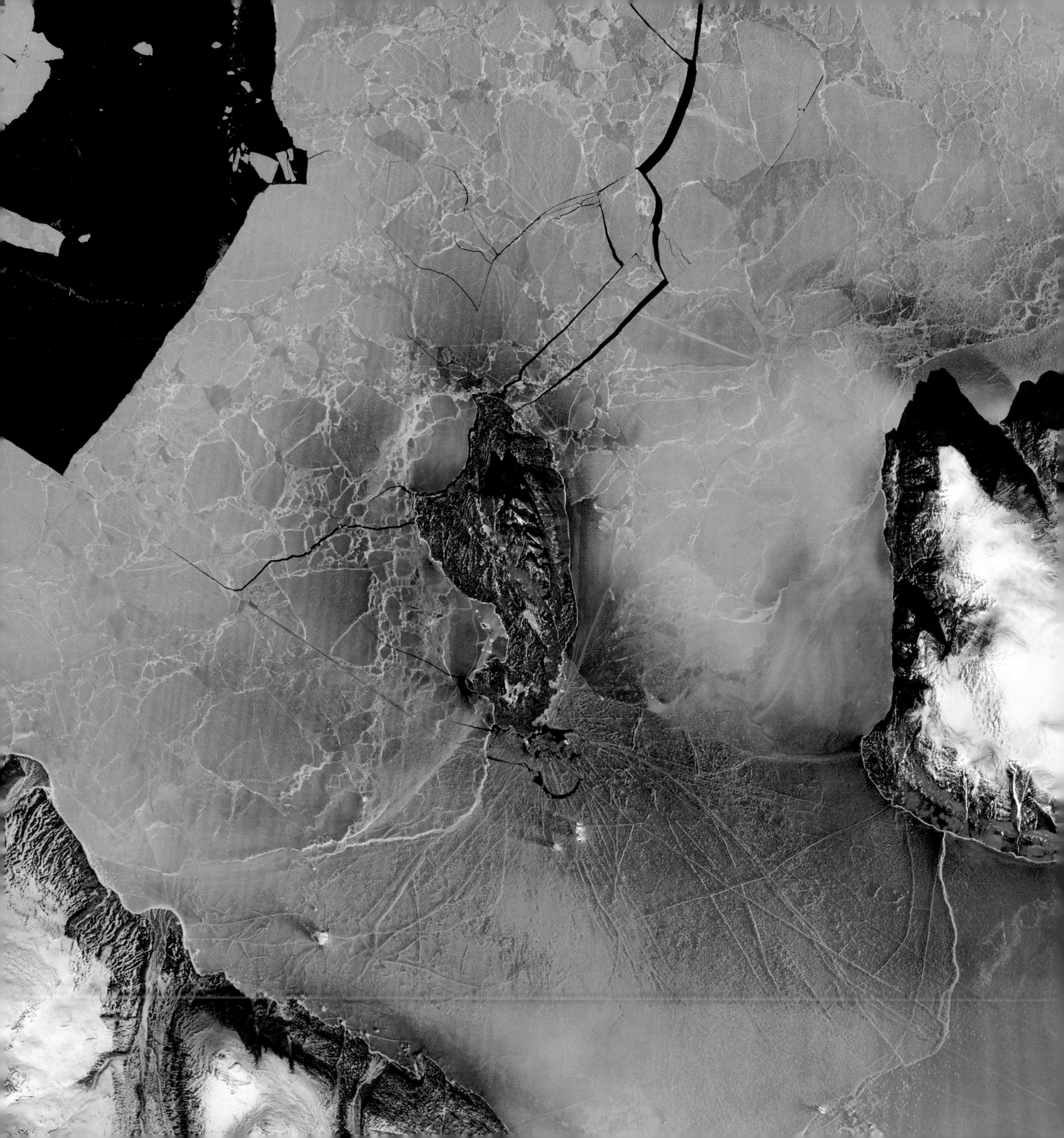

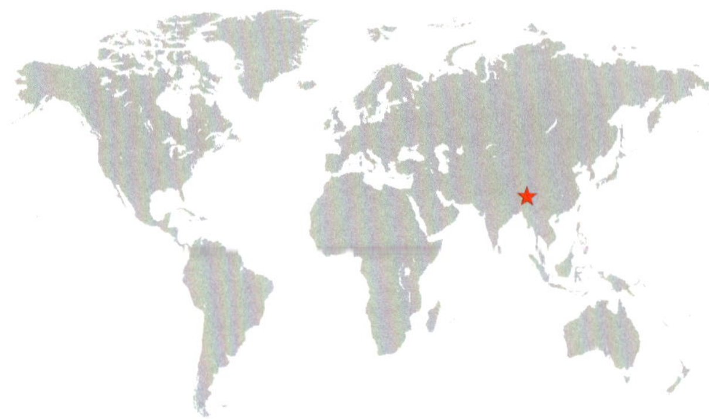

Puma Yumco
China

Several hundred lakes dot the expansive Tibetan Plateau. With the average elevation exceeding 4,500 meters (14,800 feet) above sea level, these lakes are among the highest in the world.

Recent research suggests that the number and surface area of lakes on the Tibetan Plateau has increased significantly since the 1990s.

Puma Yumco is one of the larger lakes in southern Tibet. Tuiwa, a small village along the eastern edge of the lake, is reportedly one of the highest settlements in the world. Every winter, Tuiwa villagers herd thousands of sheep across the lake's frozen surface to two small islands, where the soil is more fertile and the forage is better.

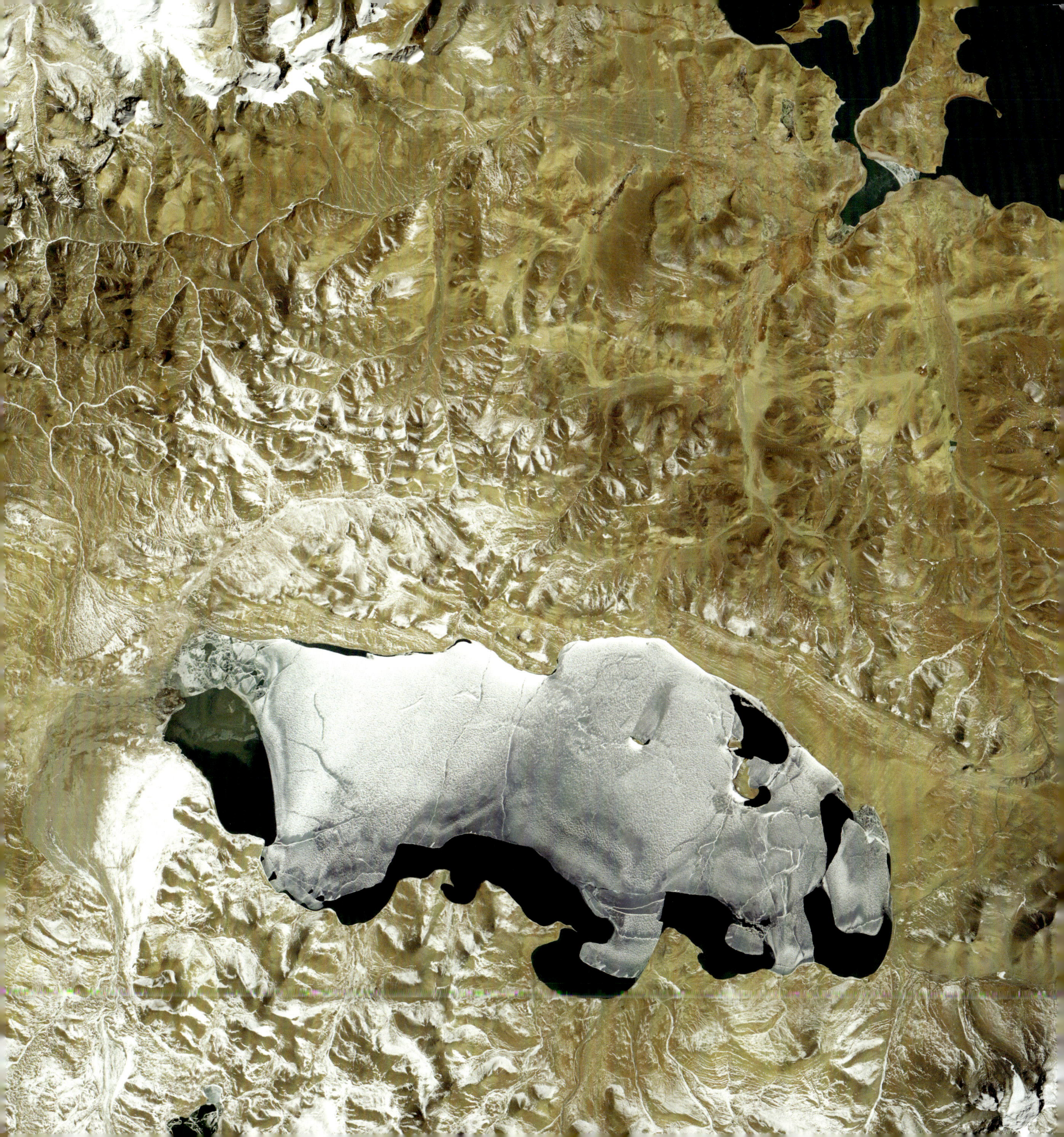

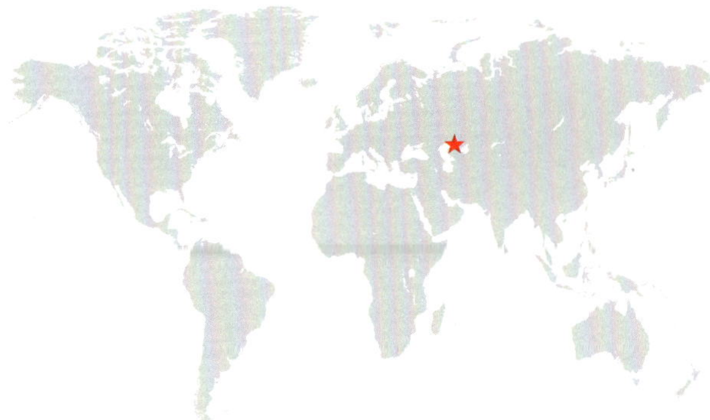

Grounded in the Caspian
Kazakhstan

A wide variety of ice forms in the Caspian Sea, which stretches from Kazakhstan to Iran. Brown areas (top left) are part of the Volga River Delta. Just offshore, a well-developed expanse of consolidated ice appears bright white. Farther offshore, a gray-white field of chunky, hummocked ice has detached and is slowly drifting around a polynya, an area of open water surrounded by sea ice. That darker patch is actually growing young, thin ice and nilas, a term that designates sea ice crust up to 10 centimeters (4 inches) in thickness.

The close-up shows nilas and a white, diamond-shaped piece of ice. It might look like this chunk is on the move, cutting a path through thinner ice. But it's more likely that the "diamond" was stuck to the sea bottom and the wind pushed ice around it.

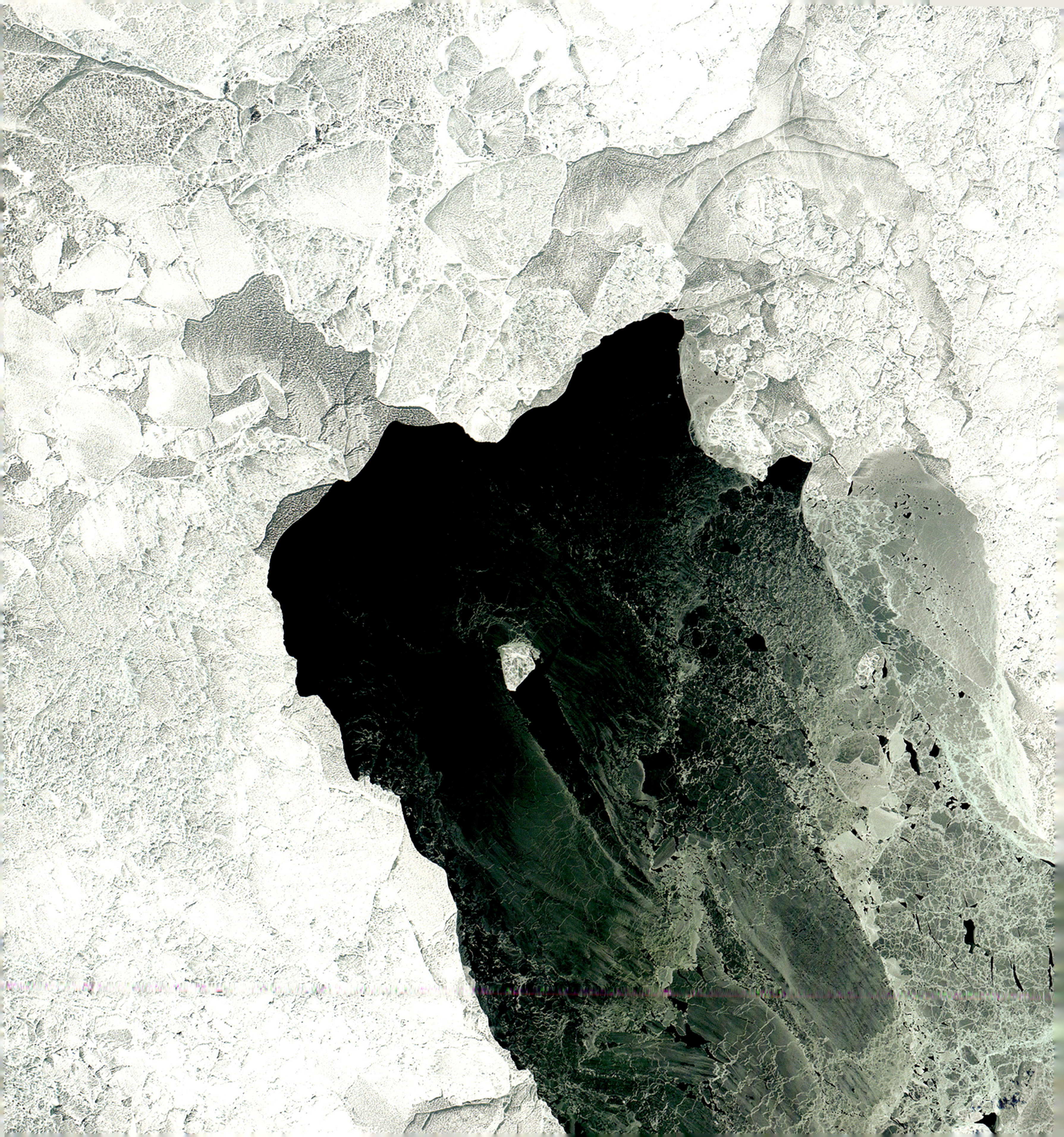

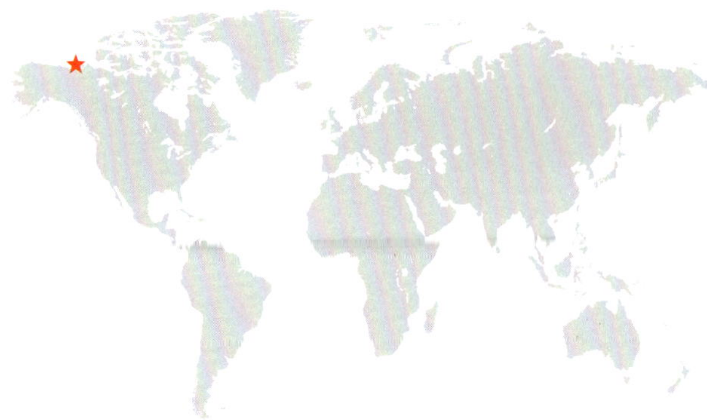

Ice-Covered Delta
Canada

In the Mackenzie River Delta of far northern Canada, snow- and ice-covered waterways stand out amid green, pine-covered land. Those frozen tributaries also become ice roads for trucks carrying supplies between the remote outposts of Inuvik and Tuktoyaktuk.

The Mackenzie River system is Canada's largest watershed, and the 10th largest water basin in the world. The river runs 4,200 kilometers (2,600 miles) from the Columbia Icefield in the Canadian Rockies to the Arctic Ocean.

Every so often, flooding from the Mackenzie River replenishes the surrounding lakes and ponds, some of which sit atop permafrost. This landscape is home to caribou, waterfowl, and a number of fish species. Also, thousands of reindeer travel through this area each year on the way to their calving grounds.

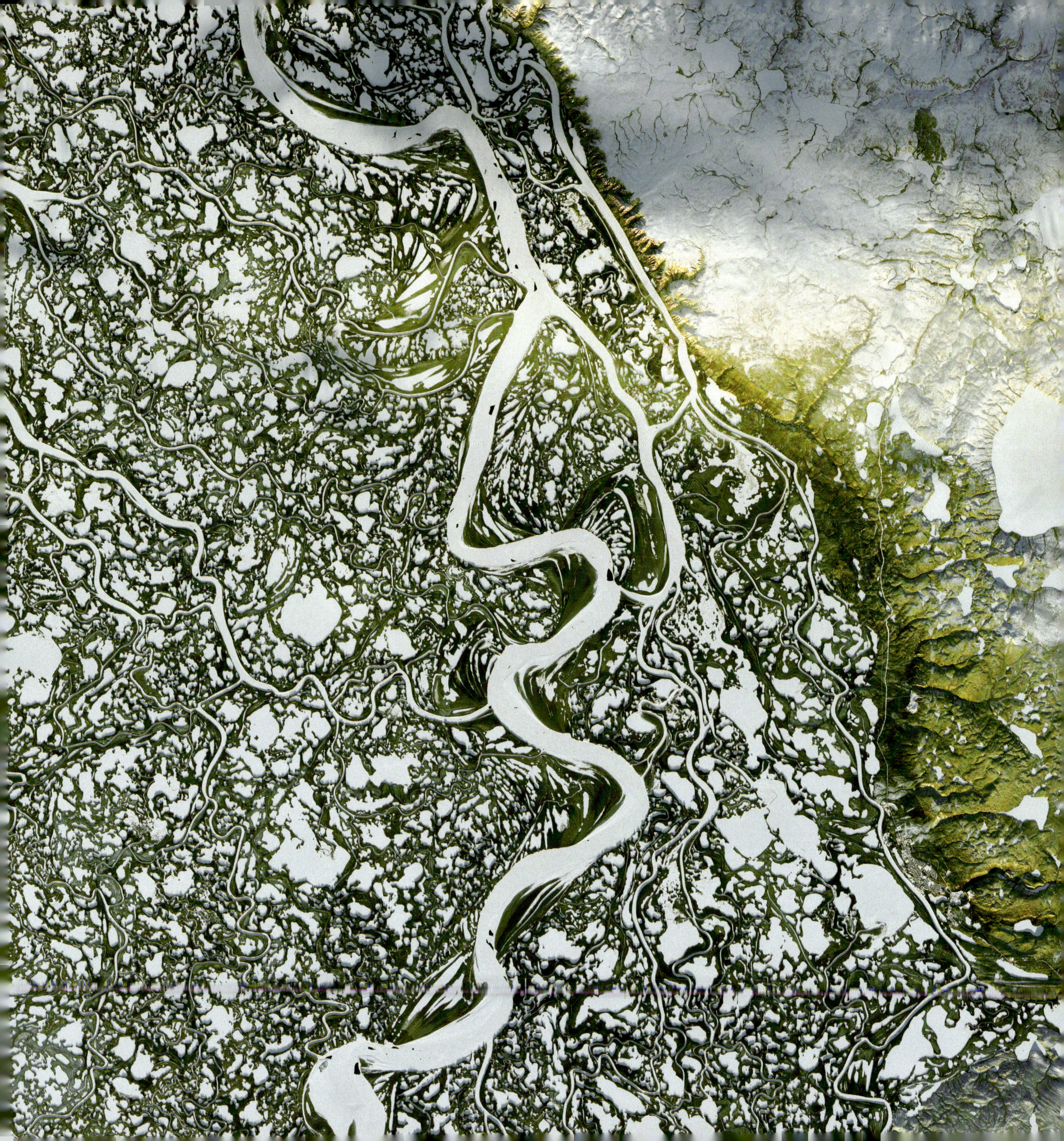

Appendix

Africa

Central and East Asia

Europe and West Asia

North America

Oceania

Oceans Atmosphere and Polar Regions

South America

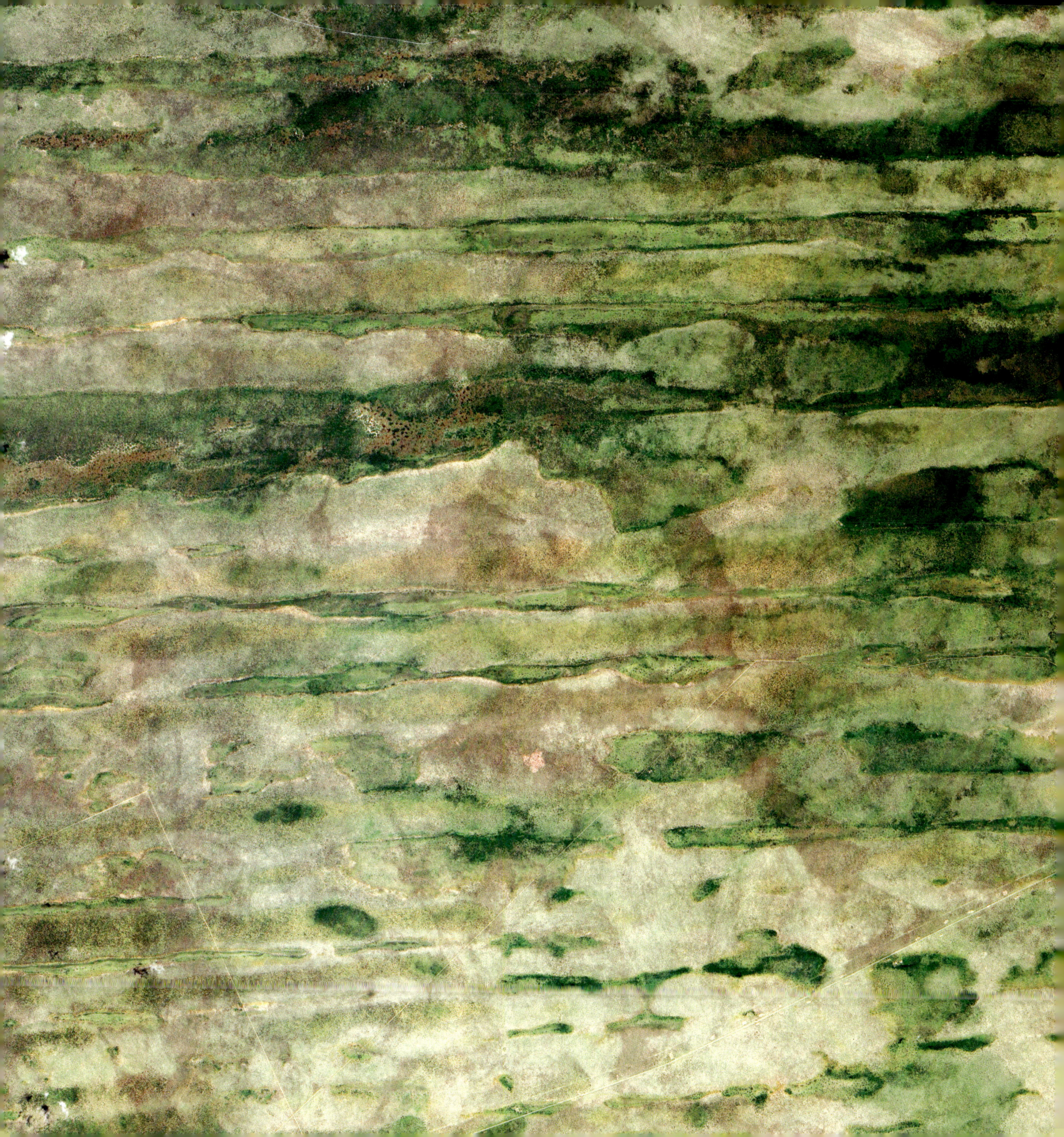

Acknowledgments

Just a few names end up on the title page of a book, but it takes an entire cast of people to bring it from idea to draft to finished product. The cast for *Earth* begins with Maxine Aldred, Andrew Cooke, Tun Hla, and Lisa Jirousek, who shepherded the words and images through design and layout. Thanks are also due to Kathryn Hansen, Pola Lem, Rebecca Lindsey, Holli Riebeek, Michon Scott, and Adam Voiland, whose reporting and writing contributions gave this book its depth. Joshua Stevens, Robert Simmon, Jesse Allen, Jeff Schmaltz, and Norman Kuring applied their strong visual sense and processing skills to make each image pop with color and texture while remaining scientifically accurate.

We owe a debt to our scientific and outreach colleagues, who keep the satellites running, the sensors sensing, and the data and imagery flowing. Every one of the images in this book is publicly available through the Internet, truly making science accessible to every citizen. The Landsat teams at the U.S. Geological Survey and NASA, the LANCE/EOSDIS MODIS Rapid Response Team, and the NASA Earth Observatory deserve extra gratitude for making our planet visible to the scientist and the layman every day.

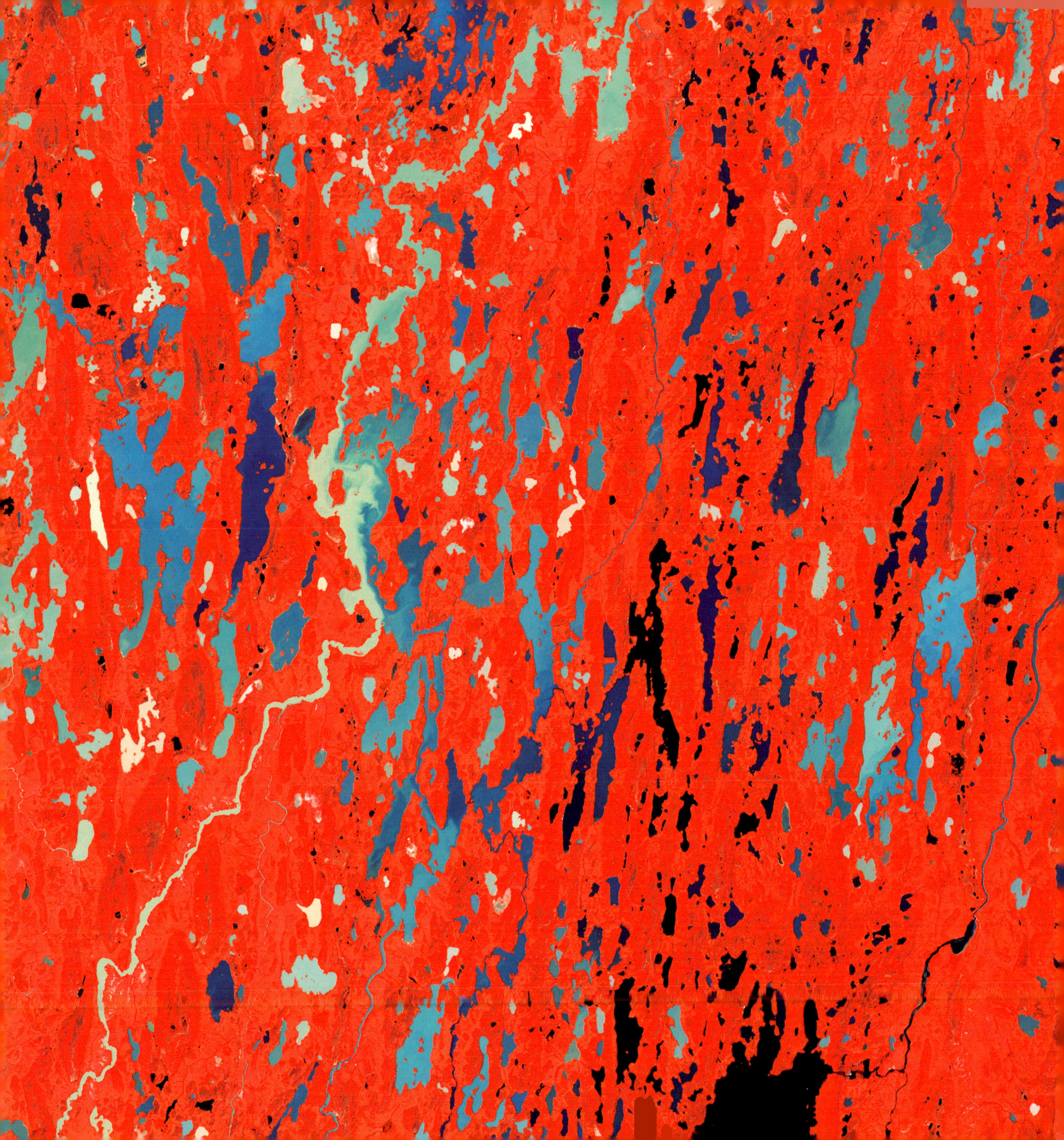

Credits

ISS Crew Earth Observations Facility: 42, 66, 70, 96, 102, 124

LANCE/EOSDIS MODIS Rapid Response Team: 4, 6, 10, 14, 24, 32

MABEL Science Team: 146

NASA Earth Observatory image/LAADS DAAC: 34, 136

NASA Earth Observatory image/NASA EO-1 data: 18, 48, 54, 60, 108, 112, 114, 128, 138, 144, 152

NASA Earth Observatory image/Suomi NPP data: 20

NASA Earth Observatory image/USGS Landsat data: 8, 12, 16, 26, 28, 36, 38, 44, 50, 52, 56, 58, 64, 76, 78, 80, 82, 88, 92, 94, 98, 100, 104, 110, 118, 120, 122, 130, 134, 140, 148, 150, 154, 156, 158

NASA Ocean Biology Processing Group: 62, 68, 74, 84

NASA/METI/ERSDAC/JAROS/Japan ASTER Science Team: 72, 116, 132

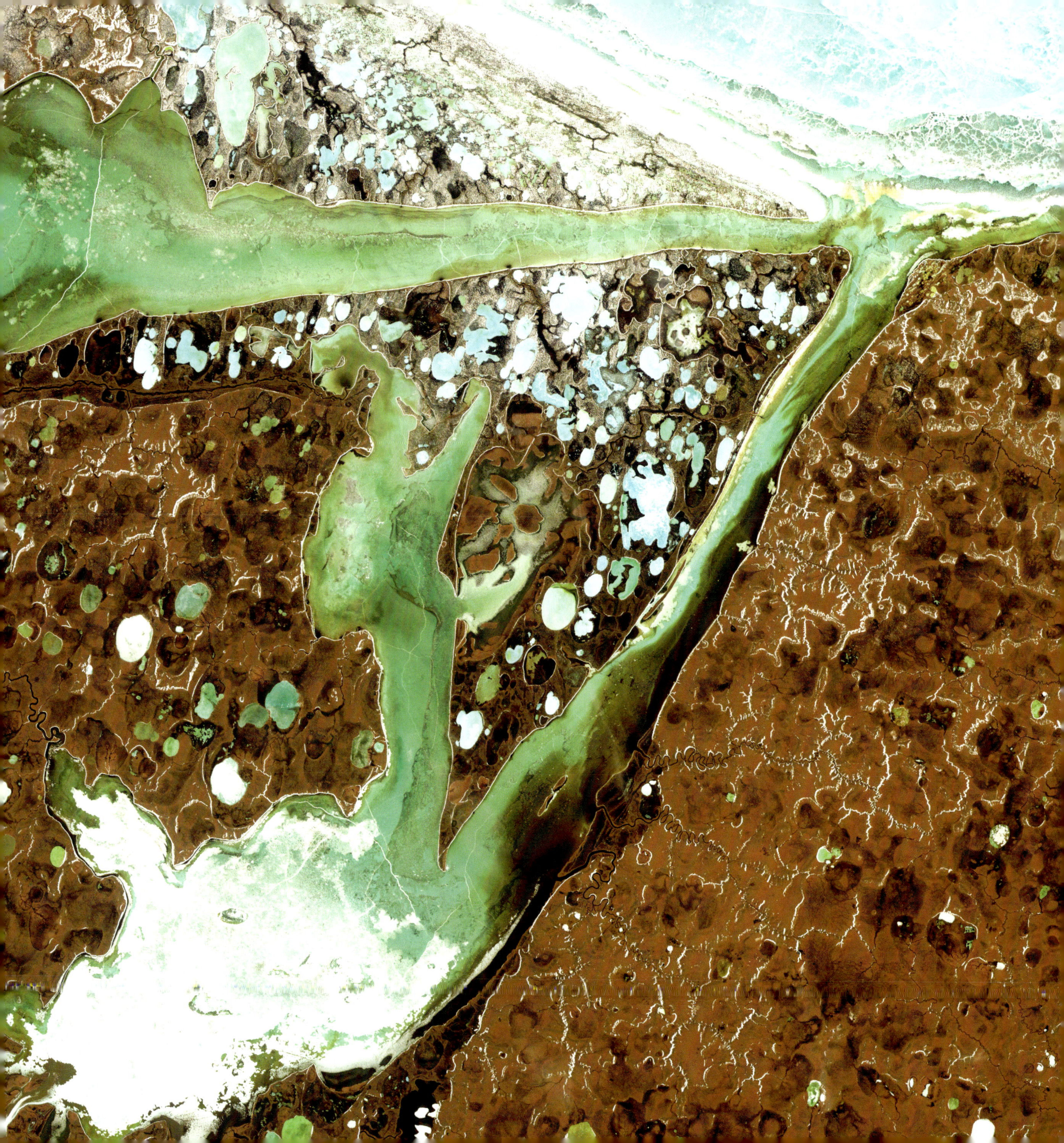

About the Authors

Michael Carlowicz is managing editor of the NASA Earth Observatory. He has written about Earth science and geophysics since 1991 for several NASA divisions, the American Geophysical Union, the Woods Hole Oceanographic Institution, and in three popular science books. He is a baseball player and fan, a longtime singer and guitarist, and the proud father of three science and engineering majors.

Kathy Carroll supports the Earth Science Division in the Science Mission Directorate at NASA Headquarters. She previously worked as a manager and organizer at for-profit and non-profit organizations and on political campaigns. She is a diehard baseball and hockey fan, and she volunteers with animal rescue organizations.

Lawrence Friedl directs the Applied Sciences Program in the Earth Science Division of NASA's Science Mission Directorate. He works to enable innovative and practical uses of data from Earth-observing satellites. He has worked at the U.S. Environmental Protection Agency and as a Space Shuttle flight controller in NASA's Mission Control Center. He and his wife have three children, and he enjoys ultimate frisbee and hiking.

Stephen Schaeberle is a graphic designer with the Communications Support Services Center at NASA Headquarters. He holds a bachelor of fine arts from the Pratt Institute, and he has received numerous awards and honors for his work and designs. He enjoys boating and fishing on the Chesapeake Bay.

Kevin Ward manages NASA's Earth Observatory Group, including the Earth Observatory, Visible Earth, NASA Earth Observations (NEO), and EONET. He holds a master's degree in library and information science and has spent more than 20 years developing Web-accessible resources in support of NASA Earth science communications. He and his wife have a son and a deep love of music.

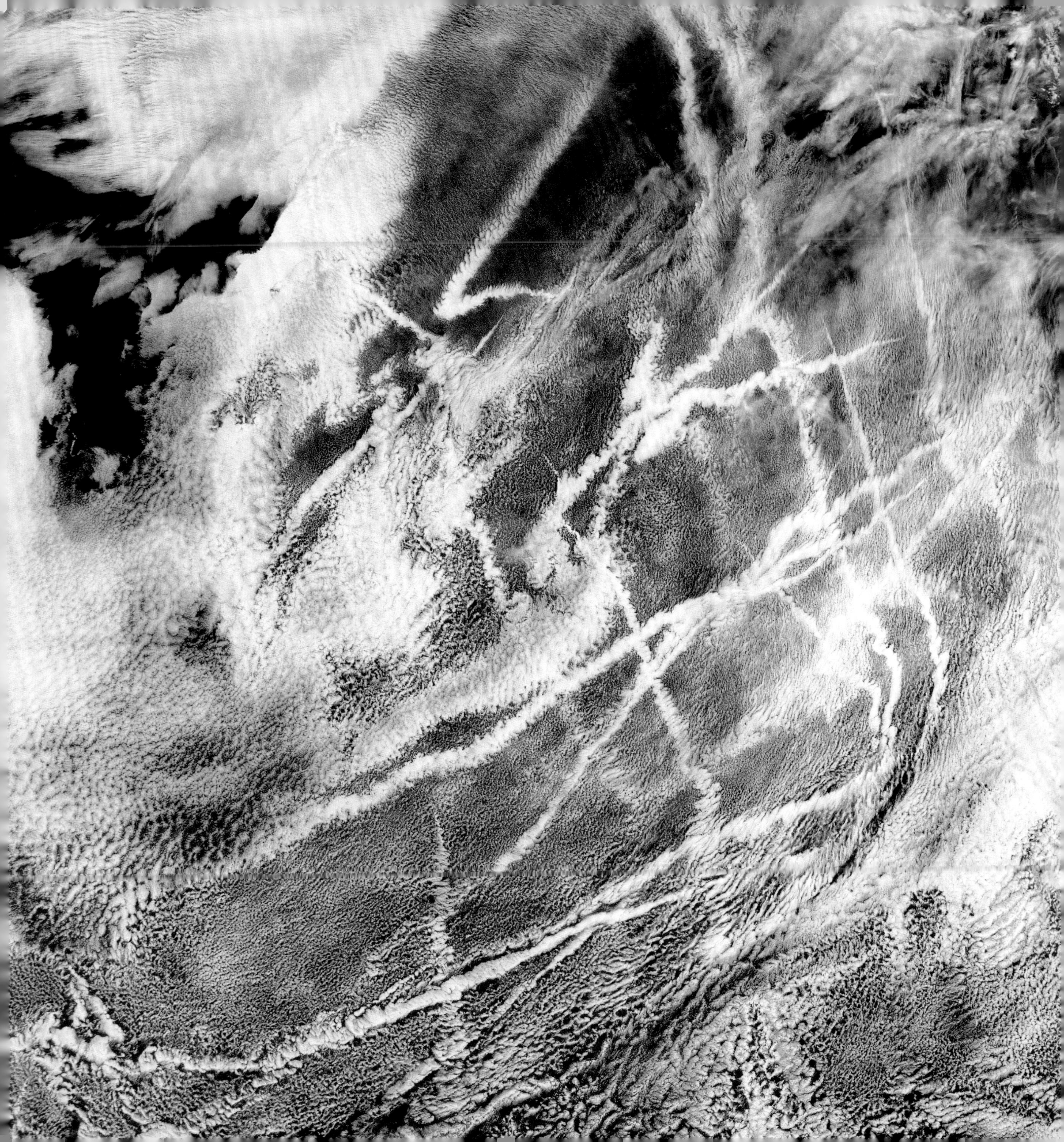